DYNAMIC PROGRAMMING AND BAYESIAN INFERENCE, CONCEPTS AND APPLICATIONS

Edited by **Mohammad Saber Fallah Nezhad**

Dynamic Programming and Bayesian Inference, Concepts and Applications
http://dx.doi.org/10.5772/57005
Edited by Mohammad Saber Fallah Nezhad

Contributors

Mark Petersen, Mmboniseni Mulaudzi, Alessandro Carbonari, Massimo Vaccarini, Alberto Giretti, Fernando Olsina, Miguel Gil-Pugliese, Claudia Notarnicola, Romina Solorza, Mohammad Saber Fallah Nezhad

Published by InTech

Janeza Trdine 9, 51000 Rijeka, Croatia

Publishing Process Manager Iva Lipovic

Technical Editor InTech DTP team

Cover InTech Design team

First published May, 2014

Printed in Croatia

Additional hard copies can be obtained from orders@intechopen.com

Dynamic Programming and Bayesian Inference, Concepts and Applications, Edited by Mohammad Saber Fallah Nezhad

p. cm.

ISBN 978-953-51-1364-5

For EU product safety concerns:
IN TECH d.o.o., Prolaz Marije Krucifikse Kozulić 3, 51000 Rijeka, Croatia,
info@intechopen.com or visit our website at intechopen.com.

Meet the editor

Mohammad Saber Fallah Nezhad is Associate Professor at Yazd University, Iran. He received his B.S., M.S. and Ph.D. degrees, all in Industrial Engineering from Sharif University of Technology, Tehran, Iran. Also he has been a visiting researcher in Karlsruhe University, Germany. He has published more than 60 research papers in national and international journals and is author of five books. Also, he has been awarded a Silver medal in 16th National Mathematics Olympiad in Iran and he has been ranked 1st in the graduate national university comprehensive exam in System Management in Iran. He has been ranked 47th among all high school graduates in Iran. His areas of interest include: reliability, quality control, quality engineering and operations research.

Contents

Preface

Dynamic programming and Bayesian inference have been both intensively and extensively developed during recent years. Because of these developments, interest in dynamic programming and Bayesian inference and their applications has greatly increased at all mathematical levels. The purpose of this book is to provide some applications of Bayesian optimization and dynamic programming.

The book focuses on a clear presentation of the main concepts and results, different models of optimization, with particular emphasis on dynamic programming and Bayesian models. It provides a description of basic material on dynamic programming and Bayesian inference, as well as more recent developments in optimization models.

Throughout the book, the chapters concentrate on concepts, rather than mathematical detail, but every effort has been made to present the key theoretical results in as precise and rigorous a manner as possible, consistent with the overall level of the book.

Prerequisites for the book are some familiarity with concept of dynamic programming, and some knowledge of Bayesian models.

Mohammad Saber Fallah Nezhad
Associate Professor of Industrial Engineering,
Yazd University, Iran

Section 1

Applications of Bayesian Inference

Chapter 1

Bayesian Networks for Supporting Model Based Predictive Control of Smart Buildings

Alessandro Carbonari, Massimo Vaccarini and
Alberto Giretti

Additional information is available at the end of the chapter

http://dx.doi.org/10.5772/58470

1. Introduction

Optimal behaviour is one the most desired features of contemporary technological systems. Challenges like secure operation, energy efficiency, and reliable performance call for the optimised behaviour of any systems that operate and interact in our living environment. The challenge in achieving optimised performances resides in the uncertainty that qualifies the environment surrounding technical systems. Whatever model drives the systems' behaviour, it must be able to face unforeseen events, to manage the vagueness of the sensing apparatus and the errors of the control devices. Bayesian statistics is one of the theoretical backgrounds that support the construction of systems which are able to act effectively inside complex environments. Bayesian statistics is grounded on the fundamental premise that all uncertainties should be represented and measured by probabilities. Then, the laws of probabilities apply to produce probabilistic inferences about any quantity, or collection of quantities, of interest. Bayesian inference can provide predictions about probability values pertaining time series or can model parameters in terms of probability distributions that represent and summarize current uncertain knowledge and beliefs. Bayesian inference uses a kind of direct causal or model-based knowledge to provide the crucial robustness needed to make the optimised behaviour of technical systems feasible in the real world [1]. Once this kind of models have been built, then theoretically sound evidence propagation algorithms are used to update the belief set about the external environment and about the system performance, on the basis of acquired evidence. This is the fundamental mechanism that drives the construction and the operation of intelligent systems based on Bayesian inference. This chapter describes a sample engineering application of this approach on a large scale. It concerns the design and the development of an intelligent building energy management system (smart BEMS) that is able to optimise the operation of the mechanical air supply systems of the Passeig De Gracia metro station in Barcelona. To the purpose of this application, predictive models were developed to support the optimal control of environmental conditions in the station, which was necessary due to the many interacting variables of the domain.

Building Energy Management Systems (BEMSs) are control systems installed in buildings for managing the building's mechanical and electrical equipment, such as ventilation, lighting, fire and security systems [2]. BEMSs consist of hardware and software components. The hardware set-up of a BEMS is typically made up of sensor-actuator networks that accurately monitor the indoor-outdoor environment and the building plants state. The software side of a BEMS consists of a number of functional layers that implement standard management functionalities like plant status monitoring, alarm management, demand driven plant management, reporting, etc. The hardware side of the commercial BEMS technology is at present a rather mature field. A number of initiatives and associations both at industrial and public level (e.g. European Building Automation and Controls Association-EU.BAC) are cooperating to develop open communication and seamless integration standards such as BACnet, KNX, LonWorks [3], and DALI [4]. The software side of commercial BEMSs is being standardised as well. Standard EN15232 provides a structured list of controls, building automation and technical building management functions that make an impact on the energy performances of buildings. Firstly, it provides a method to define the minimum requirements concerning the building automation and the building management policies, differentiated according to the level of complexity of buildings; secondly, it provides detailed methods to assess the impact of these policies on the energy performance of any given building. Nevertheless, EN15232 methods are limited to relatively simplified applications, ranging from simple homeostatic control to demand-driven and time-scheduled policies. The implementation of optimised control policies that encompass the complex weather and end-user dynamics in the energy management of buildings is still missing. The analysis of standard BEMS applications suggests that only a fraction of the available BEMS energy saving potential of each specific building is utilized by the implemented management policies, thus missing significant opportunities for reducing operating costs through better supervisory controls. Frequently, plant and building set-points follow prescribed schedules and are not optimized in response to changing dynamic conditions, including weather, internal loads, occupancy patterns, etc. Nonetheless, there are significant opportunities for optimizing control set points and modes of operation in response to dynamic forcing functions and utility rate incentives. A number of studies [5-8] have shown potential savings for optimized controls in the range of 10% to 40% of the overall cooling cost.

Model Predictive Control (MPC) [9-11] may be used to enhance BEMSs so that they can improve their control performances getting close to optimal behaviour. MPC is an advanced control technique [12] that uses the predictions of future building status, obtained by means of a model of the building's dynamics, in order to solve the problem of determining the optimal control policies. The purpose of building management is to guarantee comfort at minimum operational cost. The MPC integrated approach to building management guarantees performance over the full range of conditions which are likely to be encountered. Since the predictions

http://dx.doi.org/10.5772/58470

that serve the optimal control are obtained through model simulation of the building future states, the implementation of MPC requires the development of integrated models capable of predicting the near future behaviour of the controlled environment under specific conditions, so that the optimal solution can be sought through scenario analysis [13, 14]. Adaptive and predictive control strategies would follow from these considerations. The development of domain models that are able to drive the MPC of a complex set of systems, like the ones operating in a metro station, is not a trivial task. For example, on one side MPC models must provide accurate predictions of future states but, on the other side, they must be computationally light in order to provide predictions in a time frame compatible with the monitoring time constraints. Furthermore, MPC models must interoperate with real sensor/actuator networks that usually, for cost reasons, cannot be larger than few tenths of devices and whose deployment is constrained by a number of external factors. Nevertheless, the model accuracy must be granted despite the reduced representation of the physical model and the suboptimal selection of the parameter set. The fulfilment of such competing requirements compels the definition of a modelling framework that, by guiding the MPC modeller through a set of methodological steps, will contribute to design accurate and robust models, which are sufficiently light to be embedded in real control systems. The Bayesian inference approach, and its computational counterpart Bayesian Networks, provide the means to manage the requirements set that compels the development of MPC models.

This chapter will outline the Bayesian Network approach that was followed to develop the environmental model used to design the line 3 of "Passeig De Gracia" (PdG-L3) metro station energy control system and the main features of the hybrid modelling solution undertaken to fulfil its functional requirements. A metro station is a very complex system. It involves, among others, multi-storey underground spaces having multifaceted thermal behaviour, e.g., intricate air exchange dynamics with the outside, heat conduction with the surrounding soil and high variable internal gains due to travelling passengers and trains. Furthermore, a metro station is usually serviced by various equipment involving cooling, ventilation, safety and security, lighting, vertical transportation and horizontal passenger transfer, gates and selling machines, information and auxiliary systems. The research illustrated in this paper is one of the results of the SEAM4US project funded under EU grant n. FP7-2011-NMP-ENV-ENERGY-ICT-EeB-285408. The objective of the SEAM4US research is to develop an advanced control system for the PdG metro station in Barcelona capable of dynamically adjusting the internal environment in optimal way, based on forecasts regarding the external environment, in order to guarantee energy efficiency, comfort and comply with regulations.

2. Literature review about MPC control

As mentioned in the Introduction, the Bayesian networks developed in this chapter were used to provide forecasts about the future state of the PdG-L3 in Barcelona, given the knowledge about their current state, in order to support the application of a Model based Predictive Control (MPC) approach. In fact, MPC is an enhancement of adaptive control.

It is known that any control in buildings is targeted to minimize power consumption while keeping required comfort level and guaranteeing robustness of the solution. In order to fit these specifications, the control system must comply with several features. It must be optimal, i.e. it finds out the values of a vector of design parameters that yield optimal system performance evaluated by a so-called cost function. In addition, the control system must be adaptive, which is "a special type of nonlinear control system which can alter its parameters to adapt to a changing environment. The changes in environment can represent variations in process dynamics or changes in the characteristics of the disturbances. [...]" [15]. Robustness is also required, thus implying that the models used for designing the controller should consider all process dynamics and must be able to adapt to unknown conditions. Finally, the predictive feature is another opportunity for achieving high energy efficiencies: prediction gives the capability of taking soft control actions in advance instead of suddenly reacting to unexpected deviations from the required state, thus saving energy.

MPC works based on a model of the building dynamics and the solution of an optimization problem to determine the optimal control inputs. It takes into account the (measured) current state of the system, future weather conditions and other disturbances (e.g. internal gains), in order to control actuators (e.g. HVAC, lighting and blind systems), so that energy and money usage are minimized. At the current point in time, a heating/cooling plan is formulated for the next several hours to days, based on predictions of the upcoming weather conditions. The control action is designed by running the model of the process over a given prediction horizon and evaluating the control sequence that gives the minimum value of the cost function [16]. Based on the results from this computation, the first step of the control policy is applied to the building, setting all the HVAC components, before moving one step forward and repeating the process at the next sampling time. This receding horizon approach is what introduces feedback into the system, since the new optimal control problem solved at the next time will be a function of the new state at that point in time, and hence of any disturbances that have meanwhile acted on the building. The final result will be a trajectory of inputs and states into the future that satisfy the dynamics and constraints of the system while optimizing some given criteria.

One remarkable survey about the effectiveness of MPC was carried out by means of simulations and applied to office buildings [17]. First, the authors considered and compared a list of potential adaptive approaches, among which we cite reduction of the thermal comfort when the building is not used, widening of the room temperature comfort range, use of Indoor Air Quality controlled ventilation. Results coming from these approaches were then compared with the benefits of a model based predictive control. The building was simulated by means of a single zone, twelfth order, time discrete bilinear building model of coupled thermal, air quality and light dynamics [18, 19]. Those preliminary simulations showed that the highest energy savings were determined by predictive control, which was also the best one at reducing the number of comfort violations encountered during the process.

What makes the application of predictive control to the "Passeig de Gracia" (PdG-L3) metro station in Barcelona very meaningful is that this is the case of a large underground building where the interaction with the outdoors is very critical and it can be modelled using very complex analytical approaches and occupancy figures result someway difficult to predict. Hence, the dynamics of the station cannot be solved –and predicted– though a simplified thermal model (e.g. of statistical type); so, the reported work focuses on the key problem regarding the development of predictive models relative to complex domains, which was faced through the adoption of Bayesian Networks. In fact, they gave back a lumped representation of a number of sub-systems, involving thousands of variables.

The overall MPC control framework applied to the station is represented in Fig. 1. Inputs u to the system are the variables that can be driven by the controller (e.g. frequency that drives injector fans in the case of mechanical air supply). The outputs y are the power consumption and indicators for comfort and health that must be controlled in order to reach certain desired reference level r. The relation between inputs and outputs is also significantly affected by a set of disturbances d, such as weather, train arrival, passenger flows and fans external to the station: they cannot be manipulated but only "accounted for" by using direct measures, whenever possible, together with a disturbance model. At each control step, the prediction model receives candidate input sequences $\hat{u}$ picked out by the controller; disturbance predictions come from disturbances models $\hat{d}$, measured outputs m from PdG-L3 and the prediction model estimates the future output sequence $\hat{y}$. The optimal control sequence u^* is that one which minimizes a given cost function while complying with given constraints. Once the optimization problem has been solved, the first step u of the optimal sequence is applied as the best control action. The overall procedure is repeated at each step, thus closing the control loop. The implementation of those systems asks for the development of devices and services:

1. monitoring systems and intelligent algorithms to interpret occupant's behaviour [20];
2. high-level control systems capable of solving optimization problems in real-time;
3. accurate and fast dynamic models of buildings' behaviour and their systems, necessary to feed the high-level control systems (i.e. to generate the predicted output sequence $\hat{y}$);
4. accurate modelling of disturbances (e.g. occupancy, weather conditions etc..).

The predictive models mentioned by bullet no. 3 above were developed in the form of Bayesian Networks, because they were able to simulate the complexity of the system under analysis while keeping the computational effort manageable for real-time applications. To this aim Section 4 presents the basics on probability inference and Bayesian learning, despite the fact that this chapter cannot cover all the algorithms related to these topics. In Section 5 the indices useful to evaluate the quality of the developed networks were shown. In the same section the problem of network instantiation which includes even uncertainty is unfolded. Section 6 will report the procedure used to develop the Bayesian Networks object of this chapter, the validation of their inference capabilities and the cost function implemented by the controller to search the optimum solutions. Finally, in Section 7 the networks were wrapped within the whole predictive modelling framework and their capabilities shown through one example. Conclusions are given in Section 8.

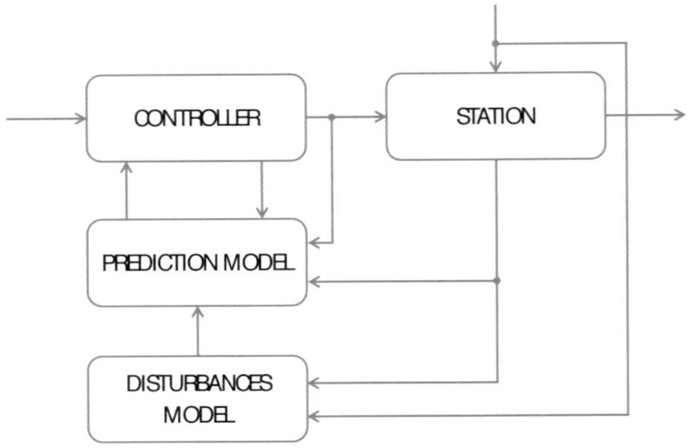

Figure 1. Predictive model based control framework defined for the metro station PdG-L3.

3. Basics of Bayesian Networks

3.1. Inference propagation

According to the framework outlined in Section 2, the current state of a building will be monitored in real-time by means of a sensor network deployed inside it. However, in order to limit the number of sensors to be installed in buildings, only a fraction of all the relevant physical variables are measured, while the remaining variables are derived from models. In other words, these models allow to estimate indirect measurements from measurements made directly by sensors. In the case study we are presenting in this Chapter, these models were developed in the form of Bayesian Networks, mainly because they are suitable to reduce complex domains into computationally manageable models. This is a key feature when computations must be performed in real-time. While numerical models (e.g. the Dymola™ model mentioned in Section 6) take hours to simulate the fluid dynamics and thermal behaviour of the PdG station, Bayesian networks make computations in a matter of seconds or minutes. Hence they are the most suitable to perform real-time predictions, provided that a reliable procedure for their development is available.

Other features typical of Bayesian Networks, which might come of advantage in these applications, are their capability of managing incomplete (e.g. one or a few data are not available because the corresponding sensors are broken) and uncertain information (e.g. if we include uncertainty in sensor measurements or if inputs are relative to forecasts of disturbance actions).

Whenever a Bayesian Network estimates indirect measurements from direct measurements, in fact it implements inference algorithms. Such inferences are computationally possible, thanks to the conditional probability relationships defined among the variables of the domain under analysis. This allows to consider just the most relevant relationships among all the variables (i.e. nodes), which are kept in the form of conditioned probabilities of the kind

$P(X_2 | X_1)$, where X_1 is a parent of X_2 (which is its child node instead) and X_2 is conditionally independent of any variable in the domain which is not its parent [21]. The "chain rule" descends from this concept, in that the joint probability of a group of variables in a domain can be determined by the knowledge of the state of just its parents, thus limiting the database required for its inference [22]:

$$P(X_1, \ldots, X_n)=\prod_{i=1}^{n} P(X_i | X_1, \ldots, X_{i-1}) \quad (1)$$

More remarkably, Bayesian Networks allow to perform inferences, in other words any node can be conditioned upon new evidences, even when they are relative to multiple variables. This feature is particularly important in case a control system must work in real-time, because in that case evidences acquired about a state variable (i.e. from sensor measurements) must be propagated to update the state of the rest of the domain. This process requires conditioning, and it might be called also probability propagation or belief updating. It is performed via a flow of information throughout the network, without limitation in the number of nodes [1]. When it is run in the MPC framework, the controller will make queries to a set of nodes belonging to the networks, whose probability distributions are computed from the state of other nodes, upon which observations (or evidences) are already available (e.g. the future state of disturbance variables and the current state of the physical domain). To this purpose the Bayes theorem is exploited when there is a need to reverse the inference. In particular, if inference is run from causes to consequences it is called predictive reasoning; otherwise, if inference is directed from consequences to causes, it is called diagnostic reasoning. Inference in Bayesian Networks is solved by complex combinations of algorithms [1]. In order to show how this works in the case of BEMs, a short example will be discussed. The first step towards the development of any Bayesian Network is defining its graphic structure, which requires all the variables of the domain to be ordered and causal relationships among them to be defined. The three elementary structures used to order variables in Bayesian Networks are: causal chains, causal tress and poly-trees (Fig. 2). Then, other more complex structures may be formed as a combination or enhancement of these elementary fragments. The computational burden would change as a consequence.

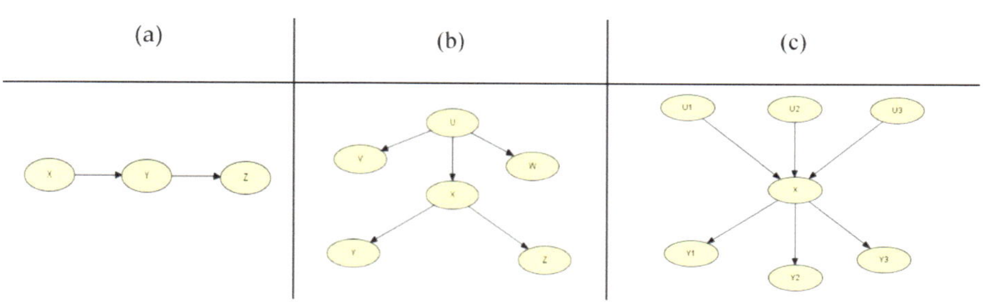

Figure 2. Graphic representation of a causal chain made up of three nodes (a), a causal tree (b) and a causal poly-tree (c).

Just to provide an example, the first structure depicted in Fig. 2-a could be useful to represent the case of sun radiation hitting and heating up the external surface of an envelope, as a consequence rising its temperature and making heat flux towards the interior. In case thermal heat gains cannot be directly measured, an alternative indirect estimation can be made: first we measure sun radiation hitting the external surface of the wall (e.g. by means of a solar pyranometer), then a model of the envelope is developed in the form of a Bayesian Network, which estimates in real-time heat gains by means of inference propagation. These models would get sun radiation intensity measured by the pyranometer as their input, then inference would infer the most likely value of internal heat gains (Fig. 3). Such an indirect estimation needs belief propagation (or probabilistic inference) based on dedicated algorithms. The notation in Fig. 3 corresponds to the notation in Fig. 2-a, provided that X stands for "sun radiation", Y stands for "wall temperature" and Z stands for "heat gain". In the latter figure four states were defined for each random variable (or event) and all the probability values were assumed as a uniform probability function (i.e. all their states are equally likely), because no learning had been done. The states of the variables were separated into intervals, according to the knowledge learned about the physical system by the model's developer, who assumed sun radiation to be limited between 0 and 400 W/m^2, wall temperature between 0 and 65°C, heat gains between 0 and 4.5 W/m^2. Once probability learning is performed as explained in the next sub-section, the network can be used to evaluate how evidence about the first variable (i.e. "sun radiation") is propagated towards the rightmost variable ("heat gain") passing through the intermediate one ("wall temperature").

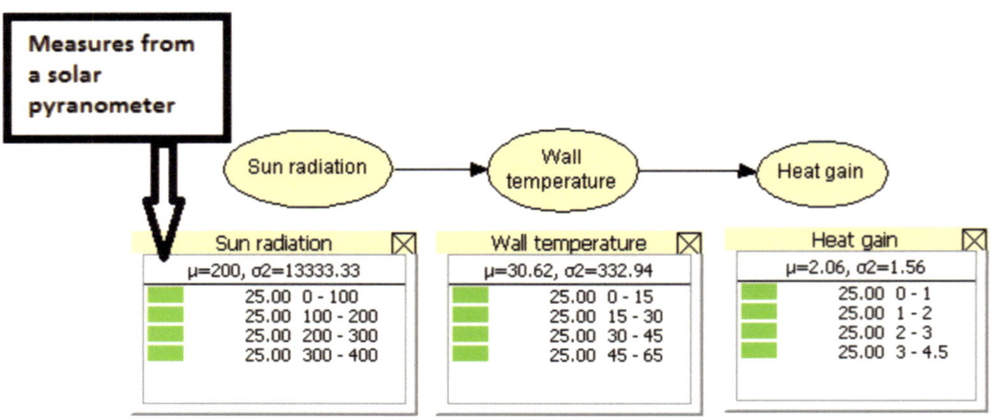

Figure 3. Example where a Bayesian chain can be used to infer indirect measurements (heat gains) from direct measurements (sun radiation intensity).

In order to make inference propagation feasible, a conditional probability table must be defined for each couple of variables in the chain. The model-based knowledge embedded in any Bayesian Network is represented graphically by a directed link between any variable X (e.g. "sun radiation") and Y (e.g. "wall temperature"), which is quantified by a fixed conditional probability matrix:

$$M_{Y|X} = P(y|x) = \begin{bmatrix} P(y_1|x_1) & P(y_2|x_1) & \dots & P(y_n|x_1) \\ \dots & \dots & \dots & \dots \\ P(y_1|x_m) & P(y_2|x_m) & \dots & P(y_n|x_m) \end{bmatrix} \quad (2)$$

where y_i and x_i are the generic states of variable nodes Y and X, respectively. In the case in Fig. 3 both X and Y might occur in one of the four possible states. Inference propagation needs to know first the probability assigned to every state of the second node Y conditioned to each state of the first node X. A more comprehensive notation is given by the form $BEL(x)$, which reflects the overall belief accorded to proposition $X=x$ by all evidence so far received, hence $BEL(x)=P(x|\xi)$. Hence this represents the set of dynamic values obtained as a result of the updated belief accorded to proposition $X=x$ once all the evidences about its parents is collected in the event ξ. In the simplest Bayesian topology, which is represented by "chains", if evidence $\xi=\{Y=y\}$ is observed, then from Bayes theorem the belief distribution of X (diagnostic reasoning) is given by:

$$BEL\ (x)=P(x \mid \xi)=\frac{P(x) \bullet P(\xi \mid x)}{P(\xi)}=\beta \bullet P(x) \bullet \lambda(x) \quad (3)$$

where the likelihood vector is given by:

$$\lambda(x)=P(\xi \mid x)=P(Y=y \mid x)=M_{y \mid x} \quad (4)$$

As a consequence, the likelihood of x is the y's column of the matrix in eq. (2). It is stored as an input in node Y, so that it can be transmitted as a message to X, thus enabling X to compute its belief distribution $BEL(x)$, having as many states as the number of rows in matrix $M_{y|x}$.

Propagation is possible even if Y is not observed directly but is supported by indirect observation $\xi=\{Z=z\}$ of a descendant Z of Y, which means the chain is of the kind in Fig. 2-a. It can still be written:

$$BEL\ (x)=P(x \mid \xi)=\beta \bullet P(x) \bullet \lambda(x) \quad (5)$$

Then, conditioning and summing upon the values of Y:

$$\lambda(x)=P(\xi \mid x)=\sum_y P(\xi \mid y,\ x) \bullet P(y \mid x)=\sum_y P(\xi \mid y) \bullet P(y \mid x)=M_{y \mid x} \bullet \lambda(y) \quad (6)$$

where the fact that Y separates X from Z was used. Due to eq. (4), $\lambda(y)=P(\xi|Y)=M_{z|y}$, so in this case it is derived from the conditional probability matrix between variables Y and Z. To the purpose of conditional updating in the chain, the state of X is irrelevant to infer the state of Z, once the state of Y is known, which might be explained by the sentence "Y d-separates X from Z" [1].

Starting from these concepts, further and more complex computations can be accomplished, so as to propagate inference throughout networks with any kind of admissible connections, but they all need the prior definition of conditional probability tables between linked variables.

3.2. Learning conditional probability tables

In the case of PdG-L3 presented in this chapter, the Bayesian Networks were built in the Hugin™ software environment. This software offers three approaches to learn the entries of conditional probability tables of the kind in eq. (2): from subjective knowledge; from analytical equations; from datasets. As the procedure presented in paragraph 6 learned conditional probability tables from datasets put together through numerical simulations, this is the case that will be considered in this sub-section. In particular, the algorithm used by Hugin™ is called "EM learning". In Fig. 4 we depicted two examples of conditional probability tables required by Hugin™ in order to perform the inference propagations explained in sub-section 4.1. The left sided one (Fig. 4-a) is the table needed to define how node *Y* of the network in Fig. 2-a is conditionally dependent to node *X*; while the right sided one (Fig. 4-b) is the table needed to define the likelihood of each state of the node *X* with respect to the joint combinations of the states of nodes *U1, U2* and *U3* (i.e. *X*'s parents). It is worth remarking that in this second case the number of columns is equal to the number of combinations of states of the nodes which are parents of *X*.

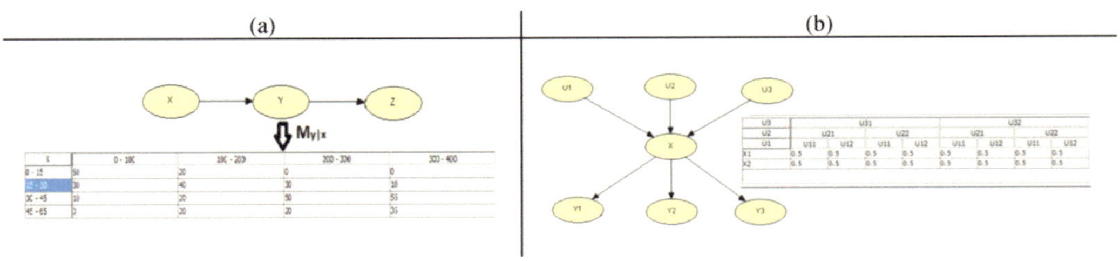

Figure 4. Conditional probability tables relative to the a node having just one parent with four states (a) and to a node having three parents with two states each (b).

The case we are considering might be described as a set of discrete variables $U=\{x_1, \ldots, x_n\}$, whose multivariate joint probability distribution can be encoded in some particular Bayesian Network structure B_s. The structure of B_s for our case study was derived from expert knowledge, hence just probabilities must be inferred from the additional knowledge provided by a random sample. Also, and according to the references in this field, it is assumed that the random sample $D=\{C_1, \ldots, C_m\}$ contains no missing data [23], which means that each case C_l consists of the observations of all the variables in *U*. Also, we can say that *D* is a random sample from B_s. Indeed, B_s may be thought as a directed acyclic graph that encodes assertions of conditional independence. In fact, it orders the variables in domain *U* such that the joint probability distribution can be estimated by means of the chain rule in eq. (1). Now, for every X_i there will be some subset $\Pi_i \subseteq \{X_1, \ldots, X_n\}$ such that X_i and $\{X_1, \ldots, X_n\}$ are conditionally independent given Π_i. That is:

http://dx.doi.org/10.5772/58470

$$P(X_i \mid X_1, \ldots, X_n) = P(X_i \mid \Pi_i) \tag{7}$$

As a consequence, a Bayesian Network is made up of a set of local conditional probability distributions and a set of assertions of conditional independence. These conditional independences can be described like in eq. (7), where the parents of X_i are grouped in the set Π_i, and are useful to "d-separate" any variable (i.e., to make it conditionally independent) from the rest of B_s.

Given B_s, let r_i be the number of states of variable X_i; and let $q_i = \prod_{x_l \in \Pi_i} r_l$ be the number of states of Π_i. Let θ_{ijk} denote the physical probability of $X_i=k$ given $\Pi_i=j$ for $i=1,..,n$, $j=1,..,q_i$, $k=1,..,r_i$. Adopting this notation, the following equivalences are valid [23]:

$$\vartheta_{ij} \equiv \cup_{i=1}^{n} \{\vartheta_{ijk}\} \quad \vartheta_{Bs} \equiv \cup_{i=1}^{n} \cup_{j=1}^{q_i} \{\vartheta_{ij}\}, \tag{8}$$

In other words, eq. (8a) states that the two notations are equivalent and represent the case where all the physical probabilities of $x_i=k$ are grouped, once any x_i in domain U is selected and the states of its parents is fixed at any $\Pi_i = j$. As a consequence, eq. (8b) represents all the physical probabilities of the joint space B_s (i.e. the Bayesian Network structure), because it encompasses all the states of Π_i (i.e., parents of x_i) and all the variables x_i in domain U. Where "$\cup \ldots$" stands for the union of all the states represented by that expression. The probability distribution of child nodes must be described by a probability distribution, which may be then updated according to evidence acquired by its parents. The software Hugin™ uses the EM algorithm, which defines a Dirichlet distribution for each variable θ_{ij} (i.e. one distribution for any variable of B_s given its parents are in the state j) [23]:

$$p(\vartheta_{ij} \mid B_s, \xi) = c \bullet \prod_{k=1}^{r_i} \vartheta_{ijk}^{N'_{ijk}-1} \tag{9}$$

where c is a normalization constant (in the form of a gamma function), N'_{ijk} is the multinomial parameters of that distribution, limited between 0 and 1, finally ξ is the observed evidence. The Dirichlet distribution describes the probability of one of the variables x_i when it varies all over its states. One of the advantages provided by using a Dirichlet distribution is that it is completely defined by its multinomial parameters. In addition, its shape can be easily adapted to fit various probability density function. Finally, and more importantly, it can be demonstrated that the learning process is made easier in this way [23]. In fact, the values N'_{ijk} represent expert knowledge introduced in the network by the developer themselves. Then, if N_{ijk} is the number of observations in the database D, in which $x_i=k$ and $\Pi_i=j$, we are able to update that distribution by adding empirical information meant by the parameter N_{ijk}:

$$p(\vartheta_{ij} \mid B_s, \xi) = c \bullet \prod_{k=1}^{r_i} \vartheta_{ijk}^{N'_{ijk}+N_{ijk}-1} \tag{10}$$

Thanks to this algorithm, which is easily implemented by computer programs, the distribution re-adjust its shape according to data provided by available sample data. It exploits the notion of experience, that is quantitative memory which can be based both on quantitative expert judgment and past cases. The prior parameters in the first version of the network are set with a particular equivalent sample size based on expert knowledge (by tuning the values N'_{ijk}). Then subsequent data are added, starting from the given equivalent sample size, so that the two Dirichlet parameters reshape the probability density function according to observations included in the dataset, which determine the values of the set of parameters N_{ijk}. Furthermore, it can be shown that the probability that $X_i=k$ and $\Pi_i=j$ in the next case C_{m+1} to be seen in the database (i.e. expected value) is computed by means of the equation:

$$P(C_{m+1} \mid D, B_s, \xi)=\prod_{i=1}^{n}\prod_{j=1}^{q_i}\frac{N'_{ijk}+N_{ijk}}{N'_{ij}+N_{ij}} \tag{11}$$

where $N'_{ij}=\sum_{k=1}^{r_i} N'_{ijk}$, and $N_{ij}=\sum_{k=1}^{r_i} N_{ijk}$.

4. Practical implementation of the Bayesian predictor

According to what reported in sub-section 4.2, the conditional probability tables of Bayesian Networks (BNs) might be defined from a combination of a priori knowledge about the involved physical processes and a dataset of collected measures. Part of the dataset is usually used for probability estimation, while the remaining part is usually used for validating the BN. Such validation may be done according to the following steps:

1. instantiation of the BN with collected evidences;
2. evidence propagation through the Bayesian reasoning engine;
3. evaluation of estimation/prediction performance by computing a proper performance index.

Once each instantaneous index is computed, a global performance index for the whole validation dataset might be computed for each node.

4.1. The metrics used for validation

The performance of a prediction model may be evaluated by using different kinds of indices. As well known, referring to prediction of a variable at a certain time i, the difference between the predicted value $\hat{X}_i$ and the actual value X_i is defined as error $E_i \triangleq \hat{X}_i - X_i$. Since the validation of a prediction model is interested just in the magnitude of the error, either its absolute value called absolute error $AE_i \triangleq |E_i|$ or its squared error $SE_i \triangleq E_i^2$ can be used. Percentage error is here defined as the ratio (%) between the error and the actual value of the variable: $PE_i \triangleq 100 \cdot E_i / X_i$. In order to have a global performance index to be evaluated over the whole validation dataset made up of K samples, these instantaneous indices must be combined into global indices. Therefore, the mean absolute error is defined as:

$$MAE \triangleq \frac{1}{K}\sum_{1}^{K} |\hat{X}_i - X_i| \tag{12}$$

and the root mean square error is:

$$RMSE \triangleq \sqrt{\frac{1}{K}\sum_{1}^{K} (\hat{X}_i - X_i)^2}. \tag{13}$$

The indices of the predicted variables are related to different physical quantities with different units. For this reason, these indices must be normalized with respect to their typical range of variation. Two more indices are defined and practically used for evaluating prediction models:

$$NMAE \triangleq \frac{\frac{1}{K}\sum_{1}^{K} |\hat{X}_i - X_i|}{|X_{max} - X_{min}|} \tag{14}$$

$$NRMSE \triangleq \frac{\sqrt{\frac{1}{K}\sum_{1}^{K} (\hat{X}_i - X_i)^2}}{|X_{max} - X_{min}|}. \tag{15}$$

ASHRAE Guideline 14-2002 [24] establishes that for calibrated simulations, the *CVRMSE* and *NMBE* of energy models shall be determined for each calibration parameter by comparing simulation-predicted data to the utility data used for calibration. The proposed indices are the coefficients of variation of the root mean square error (*CVRMSE*) and normalized mean bias error (*NMBE*). All these indices are normalized with respect to the arithmetic mean of the variable. Following this guideline, the *RMSE* has been selected as the main performance index for evaluating the accuracy of a BN. However, the mean value of a variable is a good normalization factor only if the variable is always positive (or always negative). For this reason, the range of the considered variable has been taken as a normalization factor and the *NRMSE* has been selected as final index for the design process of the BN: it includes information about both bias and variance of the error.

4.2. Instantiation of BNs for MPC

The control block in the framework depicted in Fig. 1 needs to evaluate a cost function for each candidate control policy. However, since discrete nodes are often used in BNs for describing nonlinearities, the resulting cost function may also present significant nonlinearities. One of the main problems in MPC consists in guaranteeing closed-loop stability for the controlled system. As stated in [25], stability about MPC requires some continuity assumptions. The usual approach to ensure stability is to consider the value function of the MPC cost as a candidate Lyapunov function. Even if closed-loop stability is guaranteed in some way without the continuity assumption, the absence of a continuous Lyapunov function may result in a closed-loop system that has no robustness. This implies that, whenever possible, a continuous cost function is highly recommended in MPC: it implies better stability and robustness and faster convergence of the optimization algorithm. Moreover, from a modelling point of view, the real buildings are usually continuous systems, or at least hybrid systems (i.e. mixed discrete-continuous) but never pure discrete: a discrete approximation would produce a big variance of the prediction error.

Due to these reasons, the BNs must be continuous with respect to continuous variables. When BNs contain discrete interval nodes (this is often the case in practice, since it allows to model more complex relationships between nodes), an instantiation as "*hard evidence*" (i.e. one state only of a node is assigned 100% likelihood while the other states are given 0%) produces an abrupt variation of the outputs only when passing from one interval to the next one, thus showing a discontinuous behaviour. Therefore, an instantiation must be always done as "*soft evidence*" (i.e. more than one state of a node is instantiated, in fact shaping a probability density function). The idea, with reference to Fig. 5, is to approximate each evidence with nominal value μ and estimated standard deviation σ as a Gaussian distribution $N(\mu;\sigma)$. This distribution is then approximated by an interval discrete distribution that is actually instantiated into the BN as "soft evidence". When the evidence has been propagated, the beliefs retrieved by the network are used to iterate the prediction for an arbitrary number P of steps (see Fig. 7). The final prediction sequence is then summarized as a sequence of normal continuous distributions $N(\mu_i;\sigma_i)$, $i=1, \ldots, P$.

Technically, the key challenge is approximating a continuous probability density function by means of a discrete one, as reported in the following. Given mean value and standard deviation of a normal distribution $N(\bar{\mu}_i;\bar{\sigma}_i)$, and an interval discrete random variable defined by the values $[x_0, x_1, \ldots, x_L]$ (that corresponds to the $L+1$ bounds of the L intervals), the probabilities of each interval must be found such that mean value and standard deviation of the discrete distribution are as close as possible to $\bar{\mu}$ and $\bar{\sigma}$ respectively. As well known, a normal Gaussian probability density function (pdf) takes the form $f(x)=1/\sqrt{2\pi}e^{-x^2/2}$, $-\infty<x<\infty$.

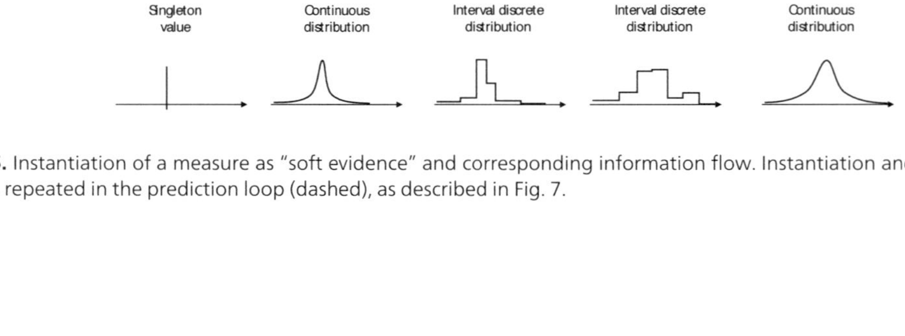

Figure 5. Instantiation of a measure as "soft evidence" and corresponding information flow. Instantiation and propagation is repeated in the prediction loop (dashed), as described in Fig. 7.

The cumulative distribution function (cdf) is then given by $\Phi(x)=1/\sqrt{2\pi}\int_{-\infty}^{x}e^{-t^2/2}dt$ that has no close form. An effective and sufficiently accurate way for approximating it is provided by [26]:

$$\Phi'(x)\approx\frac{e^{2z'}}{1+e^{2z'}},\ \text{where}\ z'=0.7988x(1+0.04417x^2) \tag{16}$$

In order to consider a general distribution with mean μ and standard deviation σ, the variable x must be replaced with $\frac{x-\mu}{\sigma}$:

$$\Phi(x)\approx\frac{e^{2z}}{1+e^{2z}},\ \text{where}\ z=0.7988\left(\frac{x-\mu}{\sigma}\right)\left(1+0.04417\left(\frac{x-\mu}{\sigma}\right)^2\right). \tag{17}$$

Given any standard deviation $\bar{\sigma}>0$, and mean value $\bar{\mu}>0$ to be approximated, the value of standard deviation must be at least greater than the standard deviation σ_{min} of the interval $[x_l;x_{l+1}]$ in which $\bar{\mu}$ falls:

$$\sigma=\max\left(\bar{\sigma},\ \sigma_{min}\right),\ \sigma_{min}=\frac{x_{l+1}-x_l}{\sqrt{12}}. \tag{18}$$

For each interval of a standard discrete node, probability is given by $q_i=\frac{q_i'}{\sum_{i=0}^{N-1}q_i'}$ where:

$$q_i'=\Phi(x_{i+1})-\Phi(x_i) \tag{19}$$

For a discrete periodic variable (like a direction) defined in the interval $[x_0;x_L]$ (that for a direction is $[0;2\pi]$), the pdf can be split into periodic intervals shifted each other by integer multiples of period (x_L-x_0):

$$[x_0+k^*(x_L-x_0);x_L+k^*(x_L-x_0)],\ k\in\mathbb{Z} \tag{20}$$

If standard deviation is sufficiently smaller than the period $\sigma\ll(x_L-x_0)$, this infinite series can be approximated by the three central intervals, that is for $k=-1,0,1$. Moreover, when $\sigma\cong(x_L-x_0)$, probability is almost equally distributed all over the interval and the approximation introduced by neglecting the stubs for $k<-1$, $k>1$ can be simply compensated by renormalizing the resulting distribution. Thus, in this case, probability of each interval is given as in the previous algorithm but the right and left stubs of the Gaussian are overlapped in order to consider periodicity of the discrete variable. For each interval of a periodic discrete node, probability is given by $p_i=\frac{p_i'}{\sum_{i=0}^{L-1}p_i'}$ where:

$$p_i' = (\Phi(x_{i+1}) - \Phi(x_i)) + (\Phi(x_{i+1} + (x_L - x_0)) - \Phi(x_i + (x_L - x_0))) + (\Phi(x_{i+1} - (x_L - x_0)) - \Phi(x_i - (x_L - x_0))) \tag{21}$$

The instantiation strategy described above was used according to the scheme in Fig. 5 for implementing the discrete approximation and smoothing the behaviour of the BNs when inputting evidences. The first Gaussian approximation of Fig. 5 has mean value equal to the measure (eventually calibrated) and standard deviation represented by the uncertainty of the measurement process, that can be determined by the data-sheet of the installed sensor. The second Gaussian approximation is trivially implemented by computing mean and standard deviation of the input distribution. Fig. 6 shows examples of Gaussian distributions approximated so as to be assigned as evidences of a Bayesian Network.

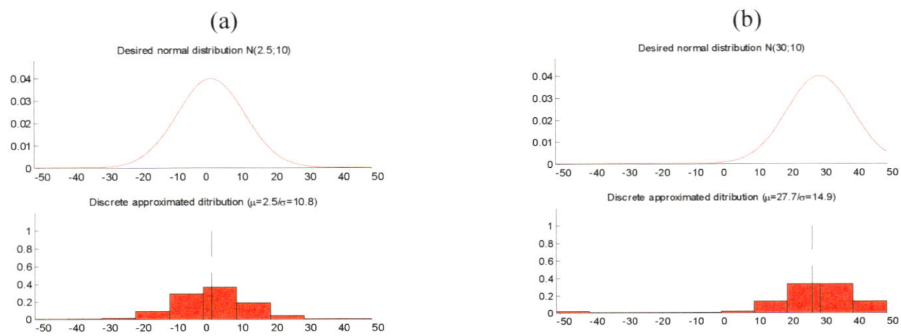

Figure 6. Comparison between the desired Gaussian distribution and the approximated one to be instantiated in an interval discrete chance node of a BN: example of a non-periodic variable (a) and of a periodic variable (b).

5. Development of the Bayesian models and cost function

5.1. The procedure

Basically, the development of both Dynamic Bayesian Networks (DBNs) and regular BNs is not an easy task and usually consists of three main phases:

1. definition of the network topology (structural learning);
2. preparation of the training set and learning of the conditional probability tables;
3. final assessment of the network.

In the case object of this chapter, the behaviour of the metro station Passeig de Gracia was first simulated through whole building analyses, which provided datasets encompassing all the possible environmental conditions, including those which are considered as not very likely, then that knowledge was transferred into Bayesian Networks. This was deemed necessary because for such complex domains eliciting expert knowledge to learn conditional probability tables is not feasible. Hence several sets of data were generated through simulations prior to the application of the EM learning process whose main features were presented in sub-section 4.2. Three datasets were generated:

- the first one was made up of randomly generated data, which means that the inputs (e.g. weather, heat gains, occupancy figures etc..) were allowed to vary without additional constraints within their range;

- the second sample, called "likely" dataset, was generated through simulations whose inputs were allowed to vary within their same ranges cited above, but their differential variations being constrained: it means that the difference between the value of each variable at the present time step and the value of the same variable at the previous time step was limited by a threshold;

- the third "typical" sample was built through simulations, whose inputs were taken from real measurements, such as real weather conditions in Barcelona, number of people passing through the station and trains recorded by occupancy surveys, internal heat gains estimated on the basis of real measurements etc..

As the learning process got information from all these samples, the resulting networks not only were expected to be able to respond to sudden variations of external actions and to consider quite unlikely events (which is the case of the first dataset) but to be very sensitive to the more likely scenarios included in the second and third datasets.

The whole building model used for running simulations was developed within the Seam4us research as a lumped parameter model in the Dymola™ simulation environment, that is based on the Modelica language [27]. Starting from a validated library for building simulation developed by the Lawrence Berkeley National Laboratory [28], a specific library for underground station was developed, which is so far in its validation phase.

However, such a model cannot be run in real-time when the controller needs to determine the best candidate control strategies, so it was reduced into the less computationally demanding form of Bayesian Networks. In order for the PdG-L3 model to predict the environmental status of the station, it was split into two Bayesian Networks, each relative to a different physical phenomenon:

- temperature prediction dynamic Network (TP-DBN), which is in the form of a DBN, because it forecasts expected temperature in the station given inputs about current and past time steps;

- air flow prediction Bayesian network (AF-BN), which is in the form of a regular BN, because it estimates variables relative to air flow in the station and energy consumption of the fans, given its current status.

Once available, the two networks were run according to the scheme outlined in Fig. 7. Here a portion of the whole process in Fig. 1 is depicted. At every iteration the controller will opportunely query the two of the networks to get future estimations about the variables relevant to select the most opportune control policy to be adopted at each running step. To this aim, the networks need to be instantiated first: the current temperature in the station's platform (PL3) and weather conditions will be provided by the permanent monitoring network installed in the station, along with candidate fan frequencies. Given these inputs, the controller is allowed to query the AF-BN in order to estimate fans consumption and air changes in the station at each time step. Such a prediction step takes a few seconds and is performed by the software Hugin™ through algorithms for belief propagation, of the kind already reported in sub-section 4.1. Then, the TP-DBN will take these variables as inputs, along with other state variables (e.g. current PL3 temperature, temperature difference between inside and outside and forecasted weather, people, train arrival etc..) in order to predict PL3 temperature at the next time step (which is typically every hour, unless a different control step is required). Again belief propagation is performed with this second network. Then, the same loop will be repeated at each iteration. This loop is based on the assumption that temperature variations between two consecutive hours is so small to be considered constant by the AF-BN without big prejudice. In other words, to the AF-BN purposes temperature at the next time step (T1) was considered equal to temperature at the current time step (T0). This assumption did not compromise the reliability of future estimations, as will be shown by the validation at the end of this sub-section. Both the networks were built following the same methodology:

1. first structural learning: it was determined first by the a-priori knowledge from the researchers, who ordered and connected the variables according to the physical relationships among them; then statistical analyses were used to improve the assumed structure and discern the strongest relationships among the variables themselves;
2. improvement of the network's structure: this was carried out through analysis of its performance indices, after learning conditional probability tables from the random dataset, which was the first set of data generated by running the Dymola™ model without constraints on all the possible weather conditions and disturbance actions;
3. final refinement using the two additional datasets: adding more datasets allowed the developers to quantify even probabilistic relationships among the variables and to define the final shape of the networks;
4. final evaluation of the networks: as for the three steps above, also in this final step the indices defined in paragraph 5.1 were used to evaluate the prediction capabilities of the networks.

The first step started from the analysis of 81 variables included in the Dymola™ dataset, properly filtered and re-sampled at a 60 sec rate. Iterative cluster analyses [29] were useful to group those variables into clusters and determine those which were redundant [30], because providing the same information given by others (e.g. surface temperatures in PL3 were highly correlated and grouped in the same level, hence just one of them was kept and the remaining ones were dropped out).

This first step allowed us to cut down the number of involved variables to 39. Then, a further reduction was done, where only those variables showing the strongest dependence relationships were kept, cutting the whole number down to 25, hence finding the minimum number

http://dx.doi.org/10.5772/58470

of nodes required to be included in the Bayesian graphs. This final set was naturally grouped into two sub-clusters of variables: one of them including those related to air flow processes, and the other one including those related to the environmental temperature dynamics. Finally, the qualitative relationships among variables (represented by arcs in the networks) were worked out partly by expert knowledge, and partly using the "structural learning" tool supplied by Hugin™, which used the information included in the "random" dataset.

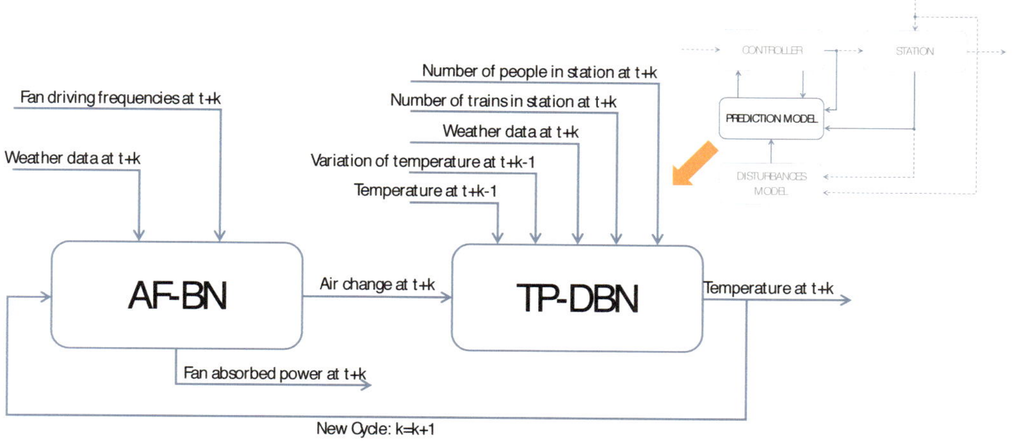

Figure 7. The basic loop of the predictive cycle adopted for PdG-L3 which involves both Bayesian Networks. It implements the prediction model in Fig.1, which starts by setting a counter at 1 and loops until the counter reaches any desired prediction horizon.

The second step was the most critical in the development process, and helped in several tasks:

- meaning and dependencies between nodes have been reviewed according to the relationships suggested by physical laws (e.g. PL3 temperature is affected by the amount of outside air supplied by ventilation and by the temperature difference between indoors and outdoors, which became its parents);
- the number of intervals for discretizing the state space pertaining to every node, which was refined through an iterative process, with the final purpose of minimizing the errors of the output variables given by the indices outlined in sub-section 5.1;
- a few links have been rearranged, in order to improve the performances of the two networks.

Fig. 8-a depicts the final structure of the dynamic Bayesian network (i.e. TP-DBN), which was used to predict PL3's temperature in PdG station in the next step (node TemPL3_p01), starting from inputs such as: forecasted number of people in the station at the next step (NPeSta_p01), forecasted internal gains supplied by trains at the next step (GaiTr_p01), current PL3's temperature (TemPL3), forecasted outdoor temperature (TOuMet_p01), forecasted air changes per hour (ACOPL3_p01) and deviation of temperature from the past time step (DTePL3). The

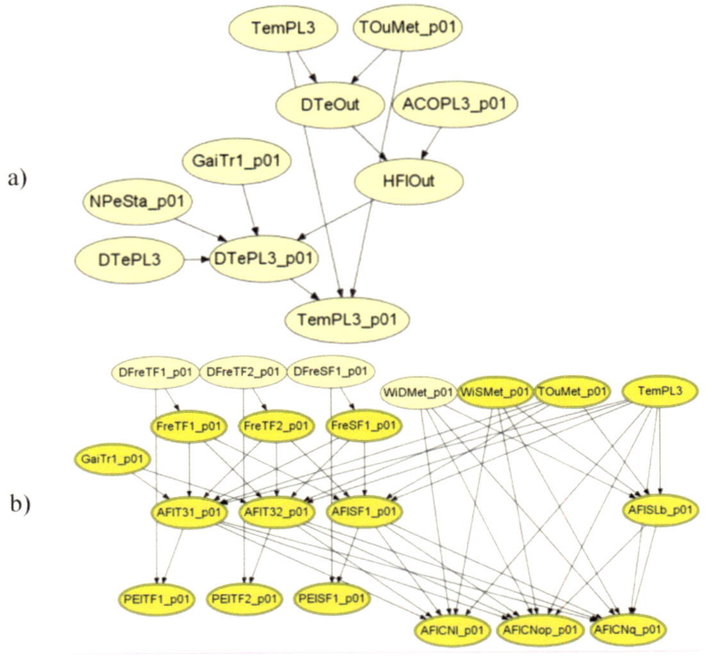

Figure 8. Predictive and dynamic Bayesian Network relative to temperature in PL3 (a) and predictive Bayesian Network relative to air flow changes in the station (b).

network's intermediate variables are useful to perform computations and simplify conditional probabilistic relationships among variables.

Similarly holds with the AF-BN network (Fig. 8-b), whose inputs are: forecasted frequencies of fans in the station and tunnels at the next time step (DFreTF1_p01, DFreTF2_p01, DFreSF1_p01), forecasted internal gains by trains (GaiTr1_p01), forecasted wind direction and speed (WiDMet_p01, WiSMet_p01), outdoor temperature (TOuMet_p01) and current temperature (TemPL3). The main outputs are the power consumption of fans – in the station (PElSF1) and in the tunnels (PElTF1, PElTF2)-and air flow rates expected across the corridors leading to PL3: AFlCNl_p01 (corridor CNl), AFlCNop_p01 (sum of corridors CNo and CNp), AFlCNq_p01 (corridoio CNq) and AFlSlb_p01 (station link). These estimated airflows are then summed up coherently to the Air Mass Balance for computing the overall air change in PL3 (ACOPL3), needed as input from the TP-DBN (Fig. 7).

During this process, whenever expert knowledge deemed more than one structure as reasonable for a certain network or fragment of it, the best one was chosen by evaluating which of them returned the highest performances according to the indices described in sub-section 5.1, after learning from the random datasets, which is the one where the widest variations and fastest dynamics of variables were considered. Even the other aforementioned amendments (e.g. rearrangement of links, optimal discretization of the networks' nodes etc..) were tested through optimization of the performances indices defined in sub-section 5.1.

The third step was aimed at performing further refinement using the "typical" and "likely" datasets, which were generated through Dymola™. Technically, that means that the EM learning algorithm was performed by adding the information included in these two datasets to the information already derived from the "random" dataset. This process allowed to include more information in the networks, mainly about those scenarios which are likely to occur more often. So it helped make the model more accurate to predict those states which are more likely to occur.

The refinement was mainly performed in terms of tuning the subdivision into intervals of the nodes and in terms of converting discrete variable into continuous variables, should they work better. The steps no. 2 and 3 required many iterations of learning, refining and validating.

Each loop was characterized by either a modification of the structure or of any network's node and conditional dependencies. Then the modified network's version was evaluated to define whether the modification had to be kept (in case it decreased the errors) or had to be rejected (in case it increased the errors), being the errors estimated as explained in section 5.1.

On the whole and just to give an example, 140 cycles were made with the TP-DBN (Tab. 1), which was useful to reduce the error from 4.98 ° to 0.72 °C (in the *RMSE* case) and from 17% to 4% (in the *NRMSE* case). The trend during the refinement process led to a continuous increment of performances, as shown in Tab. 1. In addition, 82 cycles were needed to optimize the AF-BN: for the control variable, Station Fan Power (PElSF1) *RMSE* fell down from 1858 W to 377 W, whereas *NRMSE* fell from 10.3% down to 2.3%.

In the TP-DBN all the nodes were represented by discrete variables. In the AF-BN all the variables were continuous except the following ones: frequencies of fans (DFreTF1_p01, DFreTF2_p01, DFreSF1_p01) and wind direction (WiDMet_p01).

Cycle no.	**TemPL3_p01**	
	RMSE (°C)	**NRMSE (%)**
1	4.98	17
…		
54	3.32	12
…		
98	1.81	6
…		
114	1.00	4
…		
140	0.72	4

Table 1. Gradual improvement of performances during continuous refinement of the TP-DBN network.

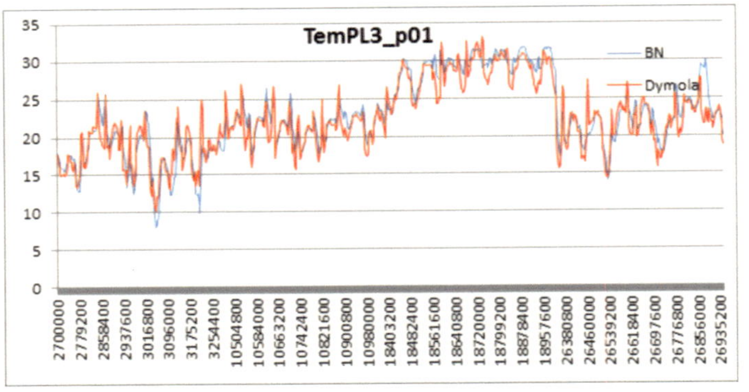

Figure 9. Qualitative comparison between the real temperature plot computed by Dymola™ and forecasts by the Bayesian Network TP-DBN.

Finally, and as fourth step, the performances of the two networks were verified also through simulations. Fig. 9 shows the good agreement between the real temperature simulated by Dymola™ in PL3 and the forecasted plot of PL3 as predicted by TP-DBN. The simulations performed by the Bayesian Networks in this case were carried out according to what already described. The input values at the first time step were instantiated as evidences taken by the Dymola model. Then, the outputs from the networks were used as inputs for the next time step in the networks and the simulations were iterated in the same way all over the period shown in the diagram. It's clear that the predictive and dynamic Bayesian network are able to accurately model the temperature plot in PL3 and to give the right inputs to the controller, in order to evaluate the best control policy.

5.2. The cost function

With reference to Fig. 1, the controller unit passes a candidate control policy to the BNs and uses resulting predictions in order to compute a cost function, which must select the best output to be used as an input in the next time step. The degrees of freedom (outputs) of the controller for PdG-L3 station are the frequencies of the station fans ($FreSF1$, $FreSF2$). The predictions that the controller queries to the Bayesian Networks are the absorbed powers of tunnel fans and station fans ($PElTF1$, $PElTF2$, $PElSF1$, $PElSF2$) and the air temperature in the platform ($TemPl3$). The future outdoor temperature ($TOuWS$) is retrieved from a weather forecast service and the air change in the platform ($ACOPl3$=amount of clean air entering the platform) is computed as a proper combination of the air flows predicted by the BNs (derived by the specific station topology):

$$\begin{aligned} ACOPl3 \triangleq\ & AFlSFa1^{+} + AFlSFa2^{+} + AFlCNl^{+} + \\ & + \left[(AFlCNop - AFlSLb + AFlCNq)^{+} - \left| AFlSLb^{-} \right| \right]^{+} \end{aligned} \quad (22)$$

The objective of MPC is to minimize the following cost function with respect to station fan frequencies:

$$J=\sum_{k=1}^{H}\alpha_{PT}\left(\frac{|PEITF1(k)+PEITF2(k)|}{\tilde{2PT}}\right)+\alpha_{PS}\left(\frac{|PEISF1(k)+PEISF2(k)|}{\tilde{2PS}}\right)+$$
$$+\alpha_{DT}\left(\frac{TOuWS(k)-TemPl3(k)}{\tilde{DT}}\right)^2+\alpha_{T}\left(\frac{\overline{TemPl3}-TemPl3(k)}{\tilde{T}}\right)^2+ \tag{23}$$
$$+\alpha_{AC}\left(\frac{\overline{ACOPl3}-ACOPl3(k)}{\tilde{AC}}\right)^2+\alpha_{DF}\left(\frac{FreSF1(k)-FreSF1(k-1)}{\tilde{DF}}\right)^2$$

Subject to constraints:

$ACOPl3(k)>ACOPl3_{Min}$

$TemPl3(k)<TemPl3_{Max}$

$FreSF_{Min}<FreSF1(k)=FreSF2(k)<FreSF_{Max}$

The variables marked with "tilde" (~) are the normalisation coefficients that corresponds to the typical values of the corresponding variable. At sampling time $t\in\mathbb{N}$, $PEITF1(k)$ represents the estimation of the value of variable $PEITF1$ at time $t+k$ $(k\in\mathbb{N})$ evaluated at time t (i.e. $PEITF1(t+k\mid t)$). The design parameters of the MPC controller are the prediction horizon H, the desired values denoted with bar notation $\overline{TemPl3}$, $\overline{ACOPl3}$, the weights of each single objective in the cost function $\alpha_{\dots}$, and the bounds of the given constraints: $ACOPl3_{Min}$, $TemPl3_{Max}$, $FreSF_{Min}$, $FreSF_{Max}$.

5.3. The networks and their estimates

The "*soft evidence*" instantiation strategy presented in section 5.2 has been implemented in a java library that wraps Hugin™ reasoning engine and allows also for other high-level functionalities, like multiple network iterations and interconnection between different networks that shares the same variables. This library is able to get the set of variables describing the current state of the station and to initialize with them the first network to be queried. Then, it is able to perform probabilistic inference by running the Hugin™ application and to extract the outputs, which will be transferred to the second network to be queried according to the procedure already shown in Fig. 7. This is done H times, if H is the desired prediction horizon.

The same functions has been also integrated in an excel spreadsheet for validating BNs. The two networks were combined according to the scheme depicted in Fig. 7 and used to simulate the predictive control. Fig. 10 shows the prediction results for one typical day (in blue), i.e. prediction horizon H=24 hours, compared with the simulation results achievable from the Dymola model (shown in red), regarding the main outputs: expected energy consumption by one of the station' fans (6a), overall airflows coming from outdoor air (6b) and future temperature (6c) plots in the platform of Line 3. In order to assess the level of accuracy of such predictions, Table 2 shows the corresponding *RMSE* and *NRMSE* relative to the variables plotted on Fig. 10. The agreement between the real plots and the estimated one is very good at each prediction time, especially for the energy consumption, that is one of the key variables influencing the cost function described in paragraph 6.2 and, as a consequence, the results of the loop managed by the controller.

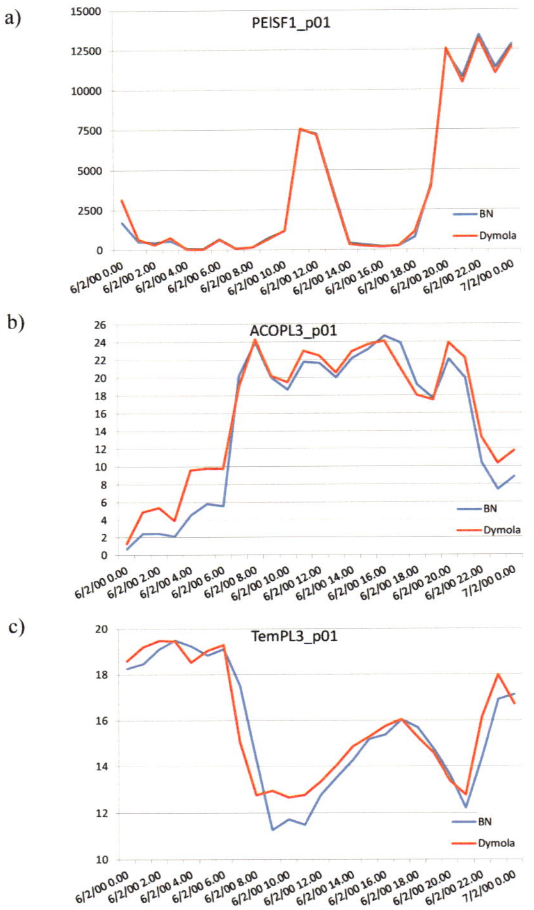

Figure 10. Comparison between real energy consumption of station fan PELSF1 and the estimated one by AF-BN (a), the real air change per hour and the estimation by AF-BN (b) and comparison between the real PL3 temperature and the estimation by TP-DBN (c).

PEISF1_p01		*ACOPL3_p01*		*TemPL3_p01*	
RMSE [W]	*NRMSE* (%)	*RMSE* [kg/s]	*NRMSE* (%)	*RMSE* [°C]	*NRMSE* (%)
326.28	0.024784	2.24	0.097566	0.91	0.13406

Table 2. Errors obtained for the whole prediction cycle with H=24 hours in the three prediction cases shown in Fig. 10.

6. The Passeig de Gracia simulator case study

In this section, we will report the implementation of a Bayesian Network predictor for the environmental control of the aforementioned test-bed given by *Passeig De Gracia* (PdG) metro station in Barcelona, Spain. The purpose of this section is to provide readers with an example regarding how Bayesian Networks can be embedded within a large MPC control framework, and how their predictions can be exploited for optimal control. The environmental control has been based on MPC because of the great complexity of the thermal and airflow dynamics in the underground environment, and of the time length of their characterising constants. The metro station underground environment is mainly characterised by huge thermal inertia caused by the terrain surrounding the station, and by relevant contributions to the indoor temperature and pollutants levels provided by air flows. Indoor air flows are determined by a number of sources that influence the station with different time frequencies and during different daytimes. The outdoor wind flow, quite frequent in a coastal station like Barcelona, affects directly the shallowest levels of the station, and determine the air pressure configuration at the interfaces of the deepest levels that may favour or impede the mechanically air supply, depending on their mutual disposal. The train transit causes the well-known piston effect. The approaching trains act as pistons compressing the air into the station platform, and leaving trains act as pulling pistons, reversing the flows. The frequency of this effect depends on the train schedule and, in our case, can be assumed about 180 sec on the average. The train piston effect causes relatively high-speed air transients with many local turbulences that affect also the platform neighbouring spaces. A third relevant source of air exchange, which is usually neglected in standard analyses, is due to the air buoyancy caused by the temperature difference between the indoor and the outdoor environments. This effect becomes substantial during the night, because of the absence of the other sources and of the greater temperature difference. Buoyancy effect causes relatively slow and laminar airflows that move air form the rail tunnels to the outside through the station's spaces.

Within this scenario, the problem of controlling the forced ventilation in an optimal way, through the fan speed, cannot be approached but with an MPC technology. Within the MPC framework, the definition of the weighting coefficients of the cost function, that maximize energy saving without compromising comfort and air quality in different operating conditions (see section 6.2), and without affecting normal operation of the station (due to safety reasons), required the development of a co-simulation architecture. Thus, the Bayesian predictor and the MPC logics has been embedded in a simulation environment that accurately reproduces the thermal and air-flow dynamics of the outdoor and indoor environments, and the trains and passenger flows. The development of the models that contribute to the simulation environment required in depth preliminary analysis by means of Finite Element Modelling, and a number of on-site surveys, that became necessary to determine the magnitude of the phenomena and to subsequently calibrate the models. Once they have been calibrated and included in the simulation architecture, the environmental models resemble the same dynamic of the measured environment, thus allowing for scenario analysis and control sizing.

6.1. The SEAM4US simulator

The Simulink (Mathworks©) architecture of the SEAM4US simulator is shown in Fig. 11. The simulator is made of four main components. The PdG Environmental model, the passenger flow simulator, the lighting control simulator and the environmental MPC.

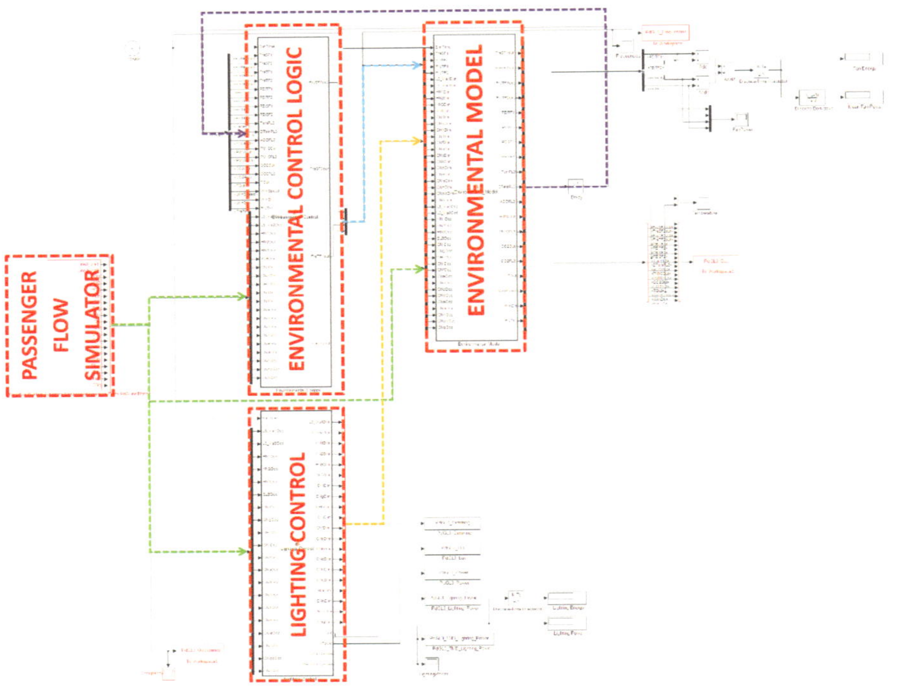

Figure 11. The Simulink SEAM4US Simulator architecture: occupancy (green), fan frequencies (blue), dimming level of lights (orange), measures (violet).

The PdG Model is the one developed in the Modelica language. To this aim, the Buildings Library 1.3 [28] has been extended in order to represent physical entities and parameters of the underground environments. At compile time the PdG environmental model results in a matrix with tenths of thousands of unknowns. The PdG Model is interfaced with a weather file of Barcelona that provides the hourly external weather parameters, including wind speed and directions. The PdG Environmental Model receives as inputs the passenger occupancy levels of each space of the station, the lighting level of the appliances in each space, and the fan control frequencies. It then outputs all the environmental parameters, like air temperature and humidity, the pollutants levels, and the energy consumption of the fans. These parameters are then fed back to the control logics as the basis for the next control step. In the SEAM4US simulator the large PdG Environmental model acts as the real station. In principle, this model could be used in the deployed MPC system to provide the necessary and accurate predictions of the next future in the model predictive controller. This option has revealed unpractical. The big size of the model produces a large memory footprint and requires significant computation power to get reasonable and effective simulation times. This is in contrast with the model embedding requirements, which foresees models included in relatively light computational environment due to deployment constraints and cost reasons. Furthermore, the alignment of the initial state of such a large model with the actual state of the station is very problematic in terms of computational time and of the stability of the solution. Therefore, the model that support the controller by means of predictions on the future status of the station has been derived from the PdG environmental model through a model reduction process into Bayesian Networks, as described in Section 6 of this Chapter.

The PdG has been used to produce a large set of control cases that have been analysed to find out the minimum set of parameters for an effective control of the target performances. The reduced case set has been then fed into the Bayesian Network, through EM learning algorithms, in order to get the Bayesian predictor for the MPC control shown in sub-section 5.1. The size of this predictor is small enough and its computational time short enough to suit the model embedding requirements. The statistical nature of the predictor avoids any problems concerning the estimation of the initial state.

The prediction accuracy achieved by the reduced model is good enough to get a reliable control of the station.

The passenger model simulates the passenger flows and the consequent occupancy distribution of the spaces inside the station. It is regulated by the train schedule, by the hour of the day – either a normal or a rush hour-and by the week day-either a weekend or a working day. The simulator has been developed in the Modelica language and is based on the bond graph theory. The passenger flow is simulated as a mass flow occurring among the station spaces. The mass sources are modulated by train arrivals and hourly scheduled flow rates observed from the outside. The model calibration has been carried out through observations (i.e. sampling performed by means of security cameras installed in the metro station) of the flow rates of passenger entering and exiting the trains and the station entrances at different hours of the days. The internal flow is then regulated through mass flow delays calculated on the basis of typical transit speeds.

Finally, the lighting control subsystems regulates the lighting level adaptively in relation to the occupancy level of each station ambient. It implements a reactive form of control that is driven by the visual task of each specific situation that may occur in the station ambient, either rush hour, waiting train arrival, and so on. The outcome of the lighting simulator are the dimming level of each appliance of the station. This of course influences the station environment as a thermal gain and is therefore provided to the PdG environmental model input.

A noteworthy aspect of the SEAM4US simulator is its computational arrangement. In fact, it implements a co-simulation architecture. The overall container is the Simulink system, which provides the control clock and a fixed simulation step to all the other subsystems. The passenger model and the lighting simulator are included in the Simulink framework as Functional Mock-up Units (FMU) that internally include the solver, through Functional Mock-up Interfaces (FMI). Interestingly, the FMI protocol provides the necessary interface between the different solvers, adapting the time varying simulation steps of the FMUs with the Simulink fixed one. The inclusion of the large PdG environmental model was not that easy. The Dymola model did not allow exporting the solver that is uniquely able to simulate the model within an FMU. Therefore, we have opted for a loosely coupled simulating environment, which runs the Dymola and the Simulink in parallel and synchronize them through a DDE channel. A wrapper of the Dymola model, which essentially implements an FMI, provides the necessary sync and message transfer functionalities.

6.2. How the MPC controller works

The control logics implemented in the SEAM4US simulator is based on a simple particle filtering mechanism (Fig. 12). The controller randomly generates a number of different control options that are sent to the predictor. The predictor updates the model with the control parameters and by means of Bayesian inference calculates the environment and energy consumption parameters. Then the controller ranks the predictor outcomes according to a cost function and to constraint satisfaction. The best performer is then selected and used in the next control step. In the PdG case described in section 6.1, control options are all the driving frequencies of the fans installed in the station and described by variables FreSF1 and FreSF2, as explained in section 5.2. The model predictor is made of the two Bayesian Networks AF-BN and TP-BN described in section 5.1. and depicted in Figs. 7 and 8. Once a set of outcomes (e.g. expected energy consumption, comfort parameters etc.) is had per each value of the inputs, the cost function in eq. (23) selects the best control strategy and sends it to the actuation system. This value is kept constant over one hour, after which another MPC step is run by the controller to re-adjust actuators for the next hour.

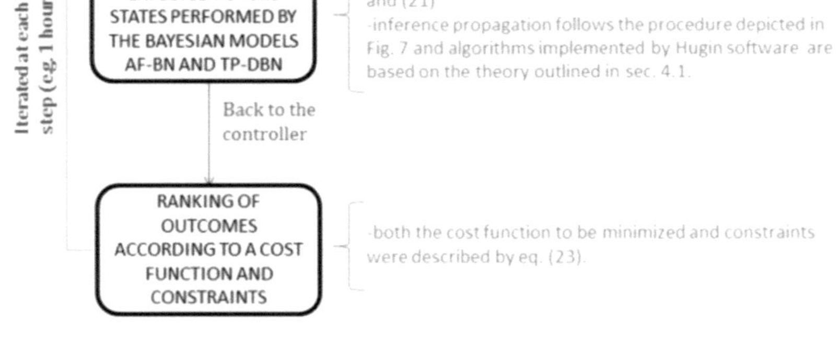

Figure 12. Flowchart showing the procedure adopted by the MPC controller for the PdG case study.

(a)

(b)

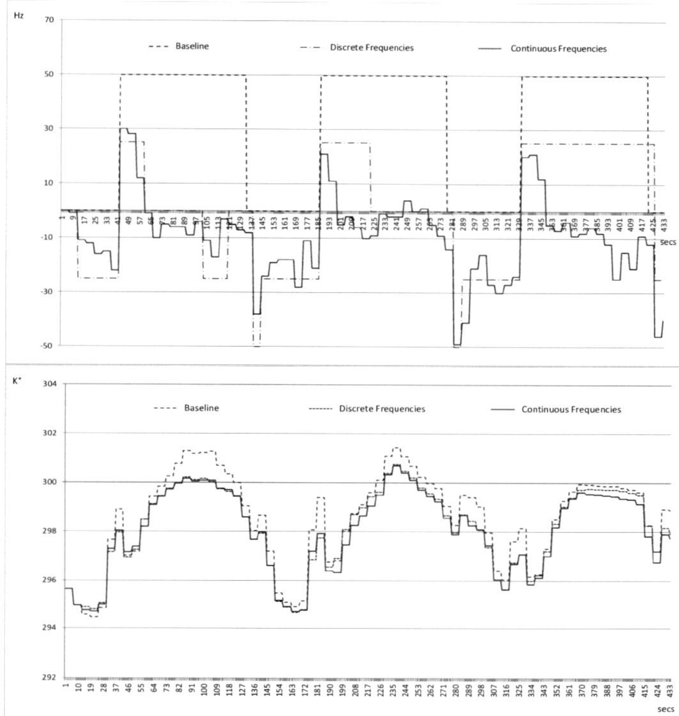

Figure 13. Some plots from the simulations in search of optimal control strategies: frequency of the fans in charge of mechanical air supply according to the control policy (top) and related air temperature plots determined by the most meaningful control strategies (bottom).

Fig. 13 shows an example of a simulation result of three days of operation, which is relative to the environmental control. The simulation time is represented along the x axis, while the y axis represents the fan frequencies in Fig. 13-a. Negative frequencies means that the fan direction is inverted (extracting air instead of supplying). Three curves are reported. The dashed curve (Baseline) depicts the current policy used for fan control, as it is actually implemented in the station, which we assumed as the baseline. The fan is driven at maximum speed for all the station opening time and it is turned off during the closure time. The second dash-dot curve represents MPC constrained to only two driving frequencies, while the third (continuous) curve is related to a continuous frequency driving. In addition to the fact that MPC control provides an energy saving rate that can rise up to 35%, it is noteworthy to realize why this happens. Comparing the baseline curve with the MPC controlled, it appears that in many cases the driving frequencies and the baseline have opposite signs. This means that in the standard baseline driving the station fans very often are opposed to the air flow induced by the external sources, and therefore contribute negatively to the air exchange and to the comfort parameters. This is reflected by the temperature curves that are slightly lower – i.e. more comfortable-for the MPC controlled environment despite the huge energy saving (Fig. 13-b). Summarizing, these results show how the effectiveness of the MPC control of complex environment relies on the power and on the flexibility of the Bayesian predictor and of the Bayesian Inference paradigm.

7. Conclusions

In this Chapter we have shown that the role of Bayesian predictors may be critical in order to implement predictive control of buildings. This kind of control is one of the most effective ones currently being developed by researchers, because it is able to smooth control actions and to trigger them in advance. However, it cannot be applied without a reliable predictor of the expected state of the controlled domain. Such a predictor must be queried in real-time, which is feasible just in case a reasonable computational effort is required.

In other words, computationally demanding software programs cannot be used to produce predictions at run time, but they can be run to generate datasets and these datasets may be used to transfer knowledge into Bayesian Networks. At this juncture, Bayesian inference may be performed: in fact, inputs by the controller (i.e. input variables describing the current state of the domain plus candidate arrays of control values) are instantiated in Bayesian Networks in the form of a set of evidences; then, inference algorithms are propagated and expected future values describing the energy and thermal state of the domain might be estimated. This procedure can be repeated thousands of times at each control step and it makes the implementation of MPC feasible.

When implemented in a real case, the results from inferences were shown to be very accurate with low deviations from the values estimated by means of more complex numerical models. In addition, our testing of the use predictive Bayesian Networks embedded in a wider MPC framework to support the ranking of concurrent control policies was successful, too. So Bayesian Networks proved to be able to solve the problem of reducing complex models into more manageable tools for performing cumbersome inferences through limited computational efforts, while getting highly accurate results.

8. List of notations

Nomenclature	Meaning
General	
BN	Bayesian Network
DBN	Dynamic Bayesian Network
TP-DBN	Temperature Prediction Dynamic Bayesian Network
AF-BN	Air Flow Prediction Bayesian Network

http://dx.doi.org/10.5772/58470

MPC	Model based Predictive Control
PdG-L3	Passeig de Gracia – Line 3 station
BEMS	Building Energy Management System
HVAC	Heat, Ventilation, Air conditioning and Cooling
Section 2	
y, $\hat{y}$	Outputs of the controlled building and their predictions respectively
u, $\hat{u}$	Inputs of the controlled building and their predictions respectively
d, $\hat{d}$	Disturbances and their predictions respectively
m	Measures from building's sensor network
Section 3	
$P(x)$	Prior probability
X, Y, Z	Random events or prepositions
$M_{y\|x}$	Conditional probability matrix
$P(y_i \| x_i)$	Any entry of the conditional probability matrix
ξ	Evidence observed before estimating future values of random variables
$BEL\ (x)$	Overall belief accorded to proposition X=x by all evidence so far received by ξ
$\lambda(x)$	Likelihood vector
β	$1/P(\xi)$
U	Domain to be modelled
B_s	Bayesian Network structure ordering a set of variables in domain U
n	Number of variables in B_s
D	Random sample
C_i	Any record of a random sample
Π_i	Any set of variables in B_s which are parents of X_i
q_i	Number of states of Π_i
r_i	Number of states of x
$_{ijk}$	Physical probability that any variable X_i in a B_s will come out with one of its states k, given its joint set of parents are in state j
c	Normalization constant in a gamma function
N	Multinomial parameter in a gamma function
Sub-section 4.1	
$\hat{X}_i$	Predicted value of any random variable X_i
X_{max}	Maximum value of variable X_i

X_{min}	Minimum value of variable X_i
K	Number of samples in a validation data set
E_i	Prediction error at time i
AE_i	Absolute error at time i
SE_i	Squared error at time i
PE_i	Percentage error at time i
MAE	Mean Absolute Error
RMSE	Root Mean Square Error
NMAE	Normalized Mean Absolute Error
NRMSE	Normalized Root Mean Square Error
CVRMSE	Coefficient of Variation of the Root Mean Square Error
NMBE	Normalized Mean Bias Error
Sub-section 4.2	
$N(\mu;\sigma)$	Normal distribution with mean μ and standard deviation σ
$N\left(\bar{\mu};\bar{\sigma}\right)$	Desired distribution with mean value $\bar{\mu}$ and standard deviation $\bar{\sigma}$
$[x_0, x_1, \ldots, x_L]$	$L+1$ bounds of the L intervals of a Interval Discrete Chance node of a BN
$f(x)$	Probability density function (pdf) of a random variable X
$\Phi(x)$	Cumulative distribution function (cdf) of a random variable X
σ_{min}	Standard deviation equals to half the interval in which a desired $\bar{\mu}$ falls
$q_i^{'}, q_i$	Probability of interval i of a non-periodic discrete node (not normalized and normalized respectively)
$p_i^{'}, p_i$	Probability of interval i of a periodic discrete node (not normalized and normalized, respectively)
Sub-section 5.1	
ACOPL3	Air changes per hour
AFlCNl, AFlCNop, AFlCNq, AFlSlb	Air flows in station's rooms named as CNl, CNop, CNq, Slb
AFlT31, AFlT32, AFlSF1, AFlSF2	Air flows in tunnels TF1 and TF2 and in station fan ducts SF1 and SF2
WiSMet	Outdoor wind speed
WiDMet	Outdoor wind direction
FreSF1, FreSF2, FreTF1, FreTF2	Frequencies that drive the two fans in the station and the two fans in the tunnels, respectively

DFreSF1, DFreSF2, DFreTF1, DFreTF2	Discretized frequencies that drive the two fans in the station and the two fans in the tunnels, respectively
PElSF1, PElSF2, PElTF1, PElTF2	Electric power absorbed by the two fans in the station and the two fans in the tunnels, respectively
TOuMet	Outdoor temperature
GaiTr1	Internal gain supplied by trains approaching on railway 1
NPeSta	Number of people in the station
TemPL3	Temperature in platform PL3
DTePL3	Variation of temperature in platform PL3
DTeOut	Variation of outdoor temperature
HFlOut	Heat flux coming from outdoor air
X_p01	Denotes one step ahead value for a variable X
Sub-section 5.2	
J	Cost function to be minimized by MPC
H	Prediction horizon
X_{Min}, X_{Max}	Minimum and maximum values allowed for a certain variable X
$\bar{X}$	Desired reference value for a certain variable X
$\tilde{X}$	Normalisation factor for a certain variable X
α	Weights of terms in cost function
X^+, X^-	Function that gets the absolute value of the positive and negative values of variableX, respectively

Acknowledgements

This work is part of the EU-funded research SEAM4US. Also, we are very grateful to our colleagues Engs. Roberta Ansuini and Sara Ruffini, who helped us develop the models.

Author details

Alessandro Carbonari*, Massimo Vaccarini and Alberto Giretti

*Address all correspondence to: alessandro.carbonari@univpm.it

Università Politecnica delle Marche, Department of Civil and Building Engineering and Architecture, Via Brecce Bianche, Ancona, Italy

References

[1] Pearl J. Probabilistic Reasoning in Intelligent Systems: Networks of Plausible Inference, 2nd ed. Morgan Kaufmann Publishers, San Mateo, California; 1998.

[2] Levermore G.J. Building Energy Management Systems: Applications to Low-energy HVAC and Natural Ventilation Control, E & FN Spon, 2000.

[3] Di Hermann M., Hansemann T., Hobner C. Building Automation: Communication systems with EIB/KNX, LON and BACnet, Springer, NY; 2009.

[4] DALI Manual, available on-line at: http://www.dali-ag.org/c/manual_gb.pdf.

[5] Daum D., Morel. N. Assessing the total energy impact of manual and optimized blind control in combination with different lighting schedules in a building simulation environment. Journal of Building Performance Simulation, 3(1), p. 1-16, 2010.

[6] Oldewurtel F., Gyalistras D., Gwerder M., Jones C., Parisio A. Increasing energy efficiency in building climate control using weather forecasts and model predictive control. In: Clima-RHEVA World Congress Proceedings, Antalya, Turkey, 2010.

[7] Oldewurtel F., Parisio A., Jones C., Morari M., D. Gyalistras D. Energy Efficient Building Climate Control using Stochastic Model Predictive Control and Weather Predictions. In: Proceedings of the American Control Conference, Baltimore, MD, 2010.

[8] Mahdavi A., Orehounig K., Pröglhöf C.A simulation-supported control scheme for natural ventilation in buildings. In: Proc. of the 11th IBPSA Conference, Glasgow, Scotland; 2009.

[9] Henze G., Kalz D. Experimental analysis of model-based predictive optimal control foractive and passive thermal storage inventory, HVAC&R, 11(2); 2005; p. 189-213.

[10] Coffey B., Haghighat F. A software framework for model predictive control with GenOpt, Energy and Buildings. 42; 2010; p.1084-1092.

[11] Clarke J. A. et al. Control in Building Energy Management Systems: The Role of Simulation, In: Proccedings of the Seventh International IBPSA Conference, Rio de Janeiro, Brazil August 13-15; 2001.

[12] Maciejowsli J. M. Predictive Control with Constraints, Prentice Hall, England; 2002.

[13] Wetter, M. A Modelica-based model library for building energy and control systems. In: Proc. of the 11th IBPSA Conf., Glasgow, Scotland; 2009.

[14] Wetter M. Modelica Library for Building Energy and Control Systems: Buildings 1.0_build2; 2001. (accessed on: Dec 09, 2011), available on-line at: https://www.modelica.org/libraries/Buildings/releases/1.0_build2

[15] McGraw-Hill, editor. Concise Encyclopedia of Engineering. The McGraw-Hill Companies, Inc.; 2002.

[16] Qin S.J., Badgwell T.A. A survey of industrial model predictive control technology. Con. Eng. Practice 2003; 11(7) 733-764; Elsevier.

[17] Gyalistras D., Gwerder M., Oldewurtel F., Jones C.N., Morari M., Lehmann B., Wirth K., Stauch V. Analysis of Energy Savings Potentials for Integrated Room Automation, 2010. Clima-RHEVA World Congress, Antalya, Turkey, May 2010.

[18] Gyalistras D., Gwerder, M. Use of weather and occupancy forecasts for optimal building climate control (OptiControl): Two years progress report. Terrestrial Systems Ecology ETH Zurich, 2010. Switzerland and Building Technologies Division, Siemens Switzerland Ltd., Zug, Switzerland.

[19] Lehmann B., Wirth K., Carl S. Modeling of buildings and building systems. In: Proceedings of the 10th REHVA World Congress Clima 2010, 9-12 May 2010, Antalya, Turkey.

[20] Rodney A. Brooks The intelligent room project: The Second International Cognitive Technology Conference, pp. 271-279; 1997.

[21] Korb K.B., Nicholson A.E. Bayesian Artificial Intelligence. Chapman & Hall/CRC Edition; 2004.

[22] Spiegelhalter D.J., Dawid A.P., Lauritzen S.L., Cowell R.G. Bayesian analysis in expert systems. Statistical Science 1993; 8(3), 219–283.

[23] Heckerman D., Bayesian Networks for Knowledge Discovery (Chapter 11), in Fayyad U.M., Piatetsky-Shapiro G., Smyth P., Uthurusamy R. Advances in Knowledge Discovery and Data Mining. MIT Press, Menlo Park/Cambridge/London; 1996.

[24] A.S.H.R.A.E. Guideline 14-2002. Measurement of Energy and Demand Savings. American Society of Heating, Ventilating and Air Conditioning Engineers, Atlanta, Georgia; 2002.

[25] Lazar M., Heemels W. P. M. H., Bemporad A., Weiland, S. Discrete-time non-smooth nonlinear MPC: Stability and robustness. In: Assessment and Future Directions of Nonlinear Model Predictive Control. Springer Berlin Heidelberg; 2007. p. 93-103.

[26] Aludaat, K. M., Alodat. M.T. A note on approximating the normal distribution function. Applied Mathematical Sciences. 2(9); 2008; p. 425-429.

[27] Mattsson S.E., Elmqvist. H. Modelica – An international effort to design the next generation modeling language. In: Boullart L., Loccufier M., Sven Erik Mattsson (editors). 7th IFAC Symposium on Computer Aided Control Systems Design, Gent, Belgium, April 1997.

[28] Wetter M. Modelica Library for Building Energy and Control Systems: Buildings 1.0_build2 (Dec 09, 2011), available at https://www.modelica.org/libraries/Buildings/releases/1.0_build2.

[29] Lawrence Hubert, Hans-Friedrich Köhn, Douglas Steinley. Cluster Analysis: A Toolbox for MATLAB. In: Roger E. Millsap, A. MaydeuOlivares. The SAGE Handbook of Quantitative Methods in Psychology. SAGE Publications Ltd; 2009. ISBN: 9781412930918.

[30] Ansuini R., Vaccarini M., Giretti A., Ruffini S. Models for the real-time control of subway stations. Proceedings. of: BS2013, Building Simulation for a Sustainable World, 13th International Conference of the International Building Performance Simulation Association, Chambery (F), August 2013.

Chapter 2

Integration of Remotely Sensed Images and Electromagnetic Models into a Bayesian Approach for Soil Moisture Content Retrieval: Methodology and Effect of Prior Information

Claudia Notarnicola and Romina Solorza

Additional information is available at the end of the chapter

http://dx.doi.org/10.5772/57562

1. Introduction

Remote sensing technologies in the microwave domain have shown the capability to detect and monitor changes related to Earth's surface variables, independently of weather conditions and sunlight.

Among these variables, soil moisture (SM) is one of the most requested ones [1]. This environmental variable is considered important to many ecological processes that occur on Earth's surface, from its relationship to climate events to its importance in terms of water availability for agricultural crops. In fact, it is considered an essential climate variable domain for the Global Earth Observation Climate (GCOS) [2]

At large scale, this biophysical variable is involved in weather and climate, influencing the rates of evaporation and transpiration. At medium-scale it influences hydrological processes such as runoff generation, erosion processes and mass movements and from the agriculture point of view determines the crops growth and irrigation needs. At small or micro-scale it has an impact on soil biogeochemical processes and water quality [3].

The ability to estimate soil moisture from satellites or airborne sensors is very attractive, especially in recent decades where the development of these technologies has taken a significant rise. This has led the possibility to have images with different spatial scales and repetition time. Despite numerous studies of moisture estimation have been developed using optical imaging, the most promising results have been obtained by using images from microwave sensors [1,4,5,6].

Satellite and aircraft remote sensing allow estimating soil moisture at large-scale, modeling the interactions between land and atmosphere, helping to model weather and climate with high accuracy [7]. In the last years many different approaches have been developed to retrieve surface soil moisture content from SAR sensors [1].

The estimation of soil moisture from SAR sensors is considered as an ill-posed problem, because many factors can contribute to the signal sensor response. The backscattering signal depends greatly on the moisture content, directly related to the dielectric constant of the soil (ε) and other factors such as soil texture, surface roughness and vegetation cover, being the latter the factors that may hinder a correct estimation of soil moisture [1].

Several studies have shown that soil moisture can be estimated from a variety of remote sensing techniques. However, only microwaves have the capability to quantitatively measure soil moisture under a variety of topography and vegetation [8]. The microwave remote sensing has demonstrated the ability to map and monitor relative changes in soil moisture over large areas, as well as the opportunity to measure, through inverse models, absolute values of soil moisture [1].

The sensitivity of SM in the microwave frequency is a well-known phenomenon, although it is still being studied by many research groups. Early researches conducted on the subject [9,10,11], among others, have shown that the sensors which operate at low frequencies of the electromagnetic spectrum, such as P or L band are capable of measuring soil moisture and overcome the influence of vegetation.

Currently most SAR systems on board of satellites (RADARSAT-2, COSMO-SkyMed, and TerraSAR-X) operate at C-and X-band, which are not the most suitable for the estimation of SM content. Some preliminary studies indicate the feasibility to estimate SM using this type of sensor, and specifically the new generation of X-band sensor [12]. However, working at such high frequencies involves dealing with interference effects introduced by the surface roughness, and especially vegetation coverage as part of the backscatter signal. Therefore, under these operating conditions, an estimate of the SM spatial variations is still a challenge.

Figure 1 shows the electromagnetic spectrum in the microwave region ordered according to the variable wavelength (in cm) and frequency (in GHz). In the same figure it is possible to have an overview of the major satellite missions, past, present and future, whose data have been used in numerous studies or can be used in the future to estimate SM.

The possibility of having multiple radar configurations was made possible thanks to the Envisat satellite launched by the European Space Agency (ESA) and its Advanced SAR (ASAR) sensor, operating in C-band [1]. Envisat/ASAR offered, unlike his predecessors, a great capacity in terms of coverage, incidence angles, polarizations and modes of operation, giving a great potential to improve the quality of many applications using SAR data.

Unfortunately, there are no current satellite missions in L band. ALOS, the satellite of Japan Aerospace Exploration Agency (JAXA), with PALSAR microwave sensor does not work since May 2011. At C-band, there is available data only from RADARSAT-2 of the Canadian Space Agency (CSA), because ERS-2 and Envisat from ESA stopped working in September 2011 and

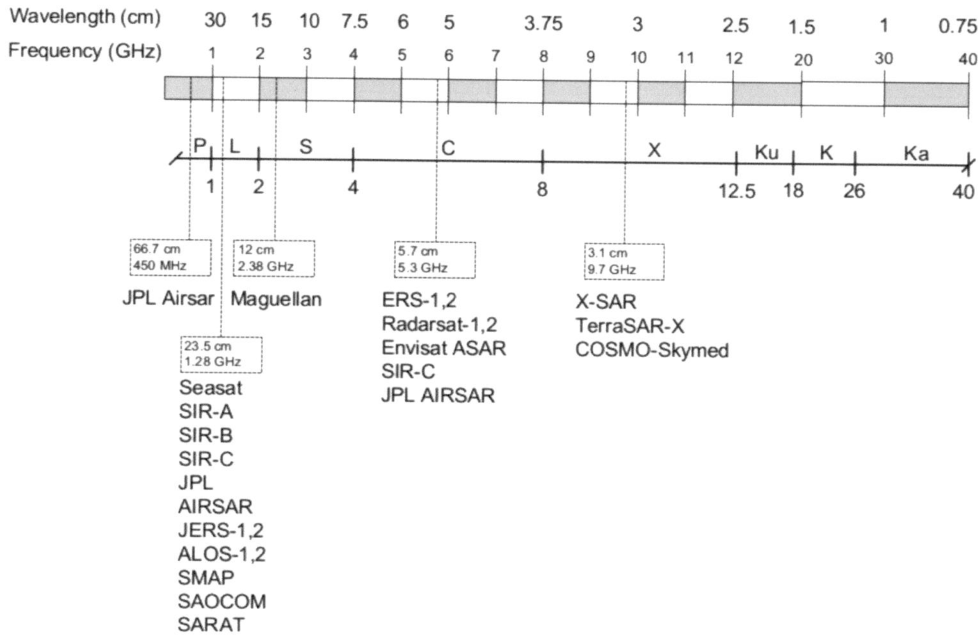

Figure 1. Main satellite missions (past, recent and future) designed in the microwave region of the electromagnetic spectrum (based on Richards, 2009).

April 2012, respectively. For the future nearby, there are expected data from planned L-band missions, such as Argentinian 1A and 1B SAOCOM, whose first launch is expected between 2014-2015; ALOS-2, which is expected to be launched in 2014. Also the SMAP active/passive satellite from the National Aeronautics and Space Administration (NASA), expected for end 2015, is very promising. SMAP will use high-resolution radar observations to disaggregate coarse resolution radiometer observations to produce SM products at 3 km resolution. The SM has been retrieved from radiometer data successfully using various sensors and platforms and these retrieval algorithms have an established heritage [7].

The most valuable information for the study of the SM has been obtained through the combination of different frequencies, polarizations and angles of incidence, as demonstrated in [11, 13,14,15]. The backscatter coefficient is highly sensitive to the micro roughness of the surface and vegetation coverage. These studies have been developed to determine the configuration of "optimal" sensor parameters, in terms of wavelength, frequency, polarization and angle of incidence to reduce interference of these factors when making an accurate estimate of SM.

In reference to specific studies, Holah et al. (2005) [16] found that an accurate estimate of SM can be achieved by using low or moderate (between 20° and 37°) incidence angles. Regarding polarizations, the most sensitive to SM are found to be HH and HV polarization while the less sensitive is VV. Li et al. (2004) [17] and Zhang et al. (2008) [18] found similar results. Furthermore Autret et al. (1989) [19] and Chen et al. (1995) [20] reported that the influence of surface roughness can be minimized by using co-polarized waves (HH/VV). Therefore, using multiple polarizations should, in theory, improve the SM estimate. The general consensus of the literature indicates that low incidence angles, long wavelengths (such as L-band) and both HH and HV polarization settings are the most appropriate sensor for an accurate SM estimate [1].

Another effective approach to mitigate the ambiguity introduced by the vegetation and roughness is to focus attention on the temporal variations through time series of radar images. In this case the basis is to assume that the average roughness characteristics and vegetation remain almost unaltered while variations in moisture content affect backscatter signal along the time [21, 22]. In this regard, methods have been developed using change detection series, as in the recent study [23] Hornacek et al. (2012) used data from the wide swath Envisat/ASAR acquisition mode as part of an evaluation of the potential of algorithms for estimating SM for Sentinel-1 mission of ESA.

In remote sensing, researchers have to deal with two problems: the direct problem and the inverse problem. The direct problem refers to the development of electromagnetic models that can correctly characterize ground backscatter coefficient by using as input the sensor parameters, such as the angle of incidence (θ_i), the wavelength (λ) and a specified polarization configuration, as well as soil parameters, such as dielectric constant and roughness.

These models provide a solid physical description of the interactions between the electromagnetic waves and the objects on the Earth's surface (e.g., bare soil or vegetation), allowing to simulate numerous experimental settings in terms of sensor configurations and soil characteristics. The generality of models is a property essential to avoid dependence on local site conditions and characteristics of the sensor, a situation that often occurs when working with evidence-based algorithms.

Once the models have been validated, it is possible to develop algorithms to invert these models and predict soil surface properties using radar observations as inputs, which is the solution to the inverse problem [24,25,26].

Numerous backscattering models have been developed in recent decades to help determine the relationship between the measured signal at the sensor and biophysical parameters, with particular emphasis on understanding the effects of surface roughness [11, 25, 27]. Considering the inversion of the direct models many approaches have been developed through numerical simulations of forward models which include Look Up Tables, Neural Networks, Bayesian approaches, and minimization techniques.

For example, the potential of some of these approaches to provide accurate maps of SM has been investigated by Pampaloni et al. (2004) [28]. They conducted a performance comparison of the three inversion algorithms using time series of Envisat ASAR cross-and co-polarized images on a farm site in Italy. The algorithms evaluated for accuracy, error rate and computational complexity were: multilayer perceptron neural network, a statistical approach based on the Bayesian theorem and iterative optimization algorithm based on the Nelder-Mead method.

Among the different methods, the Bayesian approach has been deeply investigated for its capability to provide an evaluation of the uncertainties on the variable estimates as well as the possibility to create hierarchical models with different sources of information [11, 21, 29, 30].

The objective of this research is to examine the capability and accuracy of a Bayesian approach to retrieve surface SM setting different roughness and vegetation conditions in view of an operational use of the algorithm. Several implementations of the main algorithm were designed to evaluate their different capabilities to reproduce the ground reference data. In most cases, these approaches are based on the assumption of predefined behavior of some parameters, such as vegetation and roughness, measured in situ, and then used as conditional probabilities.

The developed method has been applied to two main test sites, one located in Argentina and the second in Iowa and exploited during the SMEX′02 campaigns. The comparison over two test sites is useful to have confirmation on the behavior of the developed algorithms.

The SMEX'02 test site was one of the first exploited to test the proposed methodology that was later extended to the Argentina test site.

For this reason larger space is given in this chapter to the Argentinean test site, where SM is being deeply studied because of the near future launch of the first SAOCOM satellite. In fact, there is a particular demand of SM maps from agricultural farmers of the Pampa region for monitoring the crop status, possible evaluation of water demand and yield assessment.

2. Remote sensing data and study areas

The proposed analysis is applied to two main datasets. The first dataset derives from an experiment carried out in Argentina in view of the SAOCOM mission. The second one is located in the USA and acquired during the SMEX'02 experiments where contemporary to SAR acquisitions intensively field campaigns were carried out.

2.1. Argentinean study area

The procedure adopted here was applied to data from SARAT L Band active sensor. The SARAT SAR is an airborne sensor (figure 2) used to simulate the SAOCOM images to be analyzed in feasibility studies. The SAR Airborne instrument works in L band (λ=23cm) and is fully polarimetric.

The data set consists of field soil moisture content measurements with the corresponding backscattering coefficients at HH, HV, VH and VV polarizations and 25° incidence angle acquired with a L-band SARAT sensor. SARAT project includes an airborne sensor and an experimental agricultural site. It is part of the SAOCOM mission of Argentinean Space Agency (CONAE). The main aim of the SARAT project is to provide full polarimetric SAR images to develop and validate different applications before the launch of the first satellite SAOCOM, the SAOCOM 1A, estimated for the year 2014. The SAR instrument is installed on a Beechcraft Super King Air B-200 from the Fueza Aérea Argentina (FAA) which has a nominal range of flight altitudes between 4000 and 6000 meters above the Earth's surface, resulting in the formation of images with angles of incidence between 20° and 70°.

Figure 2. SARAT instrument and Aircraft of the FAA.

This SAR system has the same characteristics of the upcoming SAOCOM. These characteristics are described on Table 1.

Central frequency	1.3 GHz (L band)
Chirp bandwidth	39.8 MHz
Pulse duration	10 μm
Pulse Repetition Frequency	250 Hz
Swath	9 km (nominal to 4200 m of height)
Azimuth resolution	1.2 m (nominal)
Slant Range resolution	5.5 m
Spatial resolution	6 m (nominal)
Polarization	Quad-Pol (HH, HV, VH y VV)
Incidence angle	20° - 70° (nominal to 4200 m of height)
Dynamic Range	45 dB
PSLR	-25 dB
Noise equivalent σ_0	-36.9 dB

Table 1. Technical characteristics of the SARAT sensor.

SARAT project also includes a validation sites in agricultural areas. For this study, an area inside the CETT (Teófilo Tabanera Space Center of CONAE) located in Cordoba province,

http://dx.doi.org/10.5772/57562

Argentina, was selected. Its central geographic coordinates are 31°31'15.08"S-64°27'16.32"W. Figure 3 shows the location of the test site.

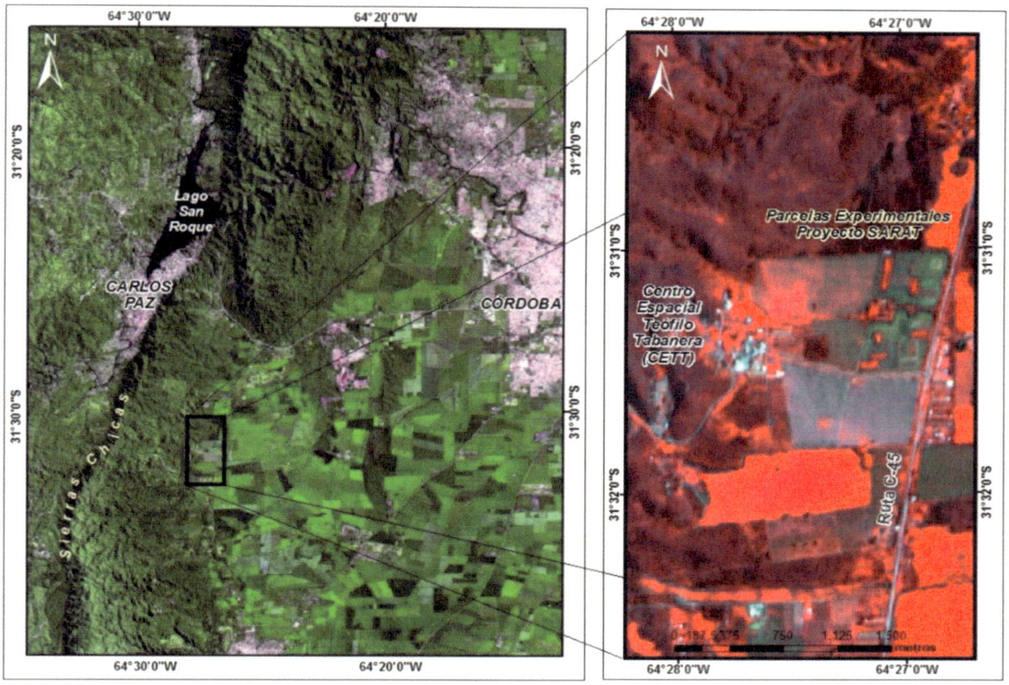

Figure 3. Location of the test site at Conae, in Argentina.

The experimental site, chosen for SM, vegetation and surface roughness measurements, has 10 fields with dimension of 50 m x 120 m which contain different kinds of crop and bare soil, as depicted in Fig. 4. All fields were intensively sampled during the SARAT acquisitions.

Plots with crops contain soybean, sunflower, corn and wheat crops. Figure 5 depicts crops stage at the moment of the SARAT acquisition time.

Some plots were left without vegetation to better investigate the interaction of microwave signal with roughness surfaces. The bare soil plots (1N, 2N, 1S and 2S) were ploughed with two roughness levels (low and high roughness) to evaluate the roughness impact on soil moisture retrieval at plot level, as shown in Fig. 6.

The roughness parameters, namely standard deviation of heights, s, and correlation length, l, were calculated as indicated in [31]. These parameters are listed in Table 2.

The SARAT images for this study (resolution: 9 m ground range) were acquired on February 2012 and all the data was provided by CONAE. Soil moisture varied between 4% and 40%, even though most plots showed medium-dry conditions.

Figure 4. Detail of the sampled plots during SARAT campaign acquisitions. N and S indicate North and South test fields.

Figure 5. Soybeans, wheat (winter development), corn and sunflower.

Figure 6. Bare plot with induced low (left) and high roughness (right).

Parameter	Symbol	High roughness value	Low roughness value
Rms height	s	3.22 cm	1.55 cm
Correlation length	l	8.17 cm	5.03 cm

Table 2. Mean roughness values inside the bare soil plots.

2.2. Iowa study area in the SMEX'02 experiment

SMEX'02 is a remote sensing experiment that was carried out in Iowa in 2002. The main site, chosen for soil moisture, vegetation and surface roughness measurements, was the Walnut Creek watershed, where 32 fields, 10 soybean and 21 corn fields, were sampled intensively [32]. The field and sensor data acquired during this experiment are particularly suitable thanks to the significant number of surveyed fields and wide range of soil conditions. The AirSAR images (resolution: 8-12 m ground range) were acquired on 1, 5, 7, 8, 9 July 2002. The five P-, L-and C-band images were processed by the AirSAR operational processor providing calibrated data sets.

3. Description of the methodology for SM estimation

The retrieval algorithm for SM is based on a Bayesian approach. Bayesian data analysis determines methods to make inference from data by using probabilities models for quantities.

The main characteristic of Bayesian methods is the explicit use of probability for quantifying uncertainty in inference based on statistical data analysis.

The process of Bayesian data analysis consists of three main steps:

- Definition of a joint probability model for all variables under evaluation;
- Calculate the posterior distribution which provides information on the unobserved quantities, given the observed data;
- Evaluation of the fit model.

Prior distributions can express our knowledge and uncertainty about the target variable. In this case the target variable could be thought as a random realization from the prior distribution.

The application of Bayesian approach implies passing from a prior distribution to a posterior distribution. Based on this concept, a relationship is expected between these two distributions [29, 30, 33]. A general feature of Bayesian inference is that the posterior distribution is centered at a point which represents a compromise between the prior information and the data. This compromise is strongly controlled by the data as the sample size increase.

A prior distribution may not have a population basis and for this reason it is desirable to have a prior which plays a minor role in the posterior distributions. These prior distributions are considered as flat, diffuse or non-informative. The rational to use such types of distributions is to let the inference being not affected by external information and based exclusively on data [34].

The proposed Bayesian approach is driven by both experimental data and theoretical electromagnetic models. The theoretical electromagnetic model has the main aim to simulate the sensor response by considering the characteristics of the soil and vegetation surface.

In order to have a better understanding of the proposed methodology, described in section 3.2, a brief description of the electromagnetic models is presented in the next section.

3.1. Electromagnetic modeling

The development of physical or theoretical models simulating direct backscatter coefficients in terms of soil attributes as dielectric constant and the surface roughness for an area of known characteristics is one of the most common approaches for SM estimation (Barrett et al., 2009). Electromagnetic models allow a direct relationship between the surface parameters and the backscattered radiation and are useful for understanding the sensitivity of the radar response to changes in these biophysical variables.

Despite its complexity, only theoretical models can produce a meaningful understanding of the interaction between electromagnetic waves and the Earth's surface. However, exact solutions of the equations that govern the rough surface scattering are not yet available and various approach methods have been developed with different ranges of validity [10]. The standard backscatter models are known as Kirchhoff Approximation (KA), which includes the Geometrical Optics Model (GOM), the Physical Optics Model (POM) and the Small Perturbation Model (SPM). These models can be applied to specific conditions of roughness in relation to the sensor wavelength. For example, GOM is considered for very rough surfaces, POM middle roughness surfaces and SPM smooth surfaces.

The Integral Equation Model (IEM), based on the radiative transfer model, has been developed by Fung and Chen in 1992 [27]. The model unifies the KA and the SPM model, a condition that makes it applicable to a wide range of roughness conditions. The IEM requires, as inputs, sensor parameters such as polarization, frequency and incidence angle, and target parameters such as the real part of the dielectric constant, the RMS height, s, and the correlation length, l [27].

For the IEM model, the like polarized backscattering coefficients for surfaces are expressed by this formula:

$$\sigma_{pp}^{0} = \frac{k^2}{2} \exp(-2k_z^2 s^2) \sum_{k=1}^{\infty} \left| I_{pp}^{n} \right| \frac{W^{(n)}(-2k_x, 0)}{n!}, \tag{1}$$

where k is the wave number, θ is the incidence angle, k_z=$kcos\theta$, k_x=$ksen\theta$ and pp refers to the horizontal (HH) or vertical (VV) polarization state and s is the standard deviation of terrain heights. The term I_{pp}^{n} depends on these parameters, k, s and on R_H, R_V, the Fresnel reflection coefficients in horizontal and vertical polarizations. The Fresnel coefficients are strictly related to the incidence angle and the dielectric constant. The symbol $W(-2k_x,0)$ is the Fourier transform of the nth power of the surface correlation coefficient. For this analysis, an exponential correlation function has been adopted that seems to better describe the properties of natural surfaces [27].

For vegetated soils, the simple approach, based on the so-called Water Cloud Model (WCM), developed by [35] has been considered in this analysis. In this radiative transfer model, the vegetation canopy as a uniform cloud whose spherical droplets are held in place structurally by dry matter. The WCM represents the power backscattered by the whole canopy σ^0 as the incoherent sum of the contribution of the vegetation σ^0_{veg} and the contribution of the underlying soil σ^0_{soil}, which is attenuated by the vegetation layer through the vegetation transmissivity parameters τ^2. For a given incidence angle the backscatter coefficient is represented by the following expression:

$$\sigma^0 = \sigma^0_{veg} + \tau^2 \sigma^0_{soil}. \tag{2}$$

If the terms related to vegetation and incidence angle are explicitly written in more detailed way, the backscattering coefficients become:

$$\sigma^0 = A \cdot VWC^E \cos\theta \cdot (1 - \exp(-2 \cdot B \cdot VWC / \cos\theta)) + \sigma^0_{soil} \cdot \exp(-2 \cdot B \cdot VWC / \cos\theta), \tag{3}$$

where VWC is the vegetation water content (kg/m^2), θ the incidence angle, σ^0_{soil} represents the backscattering coefficient of bare soil that in this case calculated by using the IEM model, τ^2 is the two-way vegetation transmissivity with $\tau^2 = exp(-2B\ VWC/\ cos\theta)$. The parameters A, B and E depend on the canopy type and require an initial calibration phase where they have to be found in dependence of the canopy type and with the use of ground data.

In this work the model simulation enters directly in the inversion procedure. For the Bayesian approach, the simulated data are generated in order to compare them to the measured data and to create the noise probability density function (PDF) as detailed in the section devoted to this approach. For this reason, it is needed to perform a preliminary validation of the proposed model as their simulation enters directly the inversion procedure.

Calibration constant values of the WCM, namely A, B and E were taken initially from literature to take into account the effect of vegetation on the SAR signal [36]. Subsequently through a Maximum Likelihood approach they were determined to fit the data used in this work from both test sites. The application of calibration equations considers two different kind of vegetation, with respect to the sensor response: very dense vegetation (as corn and sunflower) and less dense vegetation (soybean and grass). This step includes the NDVI calculation from some SPOT and LANDSAT optical images for the Argentinean and SMEX'02 test site respectively acquired close in time to the SAR image. Then the NDVI values were transformed in VWC through empirical approach defined by Jackson et al, 2004 [37].

The free parameters are illustrated in table 3, where also the RMSE of the difference between measured and simulated backscattering coefficients are reported. Figure 7 depicts the comparison between simulated and measured backscattering coefficients.

Model	A	B	E	RMSE
Soya	0.00119	0.03	0.634	1.7 dB
Corn	0.2	0.003	2.2	2.6 dB

Table 3. Calibration parameters for simulation of L band backscattering coefficients with the Water Cloud Model with distinction between soybean and corn types.

3.2. Bayesian approach for SM estimation

The objective of this research is to examine the capability and accuracy of a Bayesian approach to retrieve surface soil moisture under different assumptions for prior information on roughness and vegetation conditions in view of an operational use of the algorithm.

In the Bayesian approach, the scope is to infer biophysical parameters (e.g. soil moisture), from a set of backscattering responses measured by the sensor. The proposed algorithm is based on experimental data and theoretical models. The problem of having a few amounts of experimental data to build a reliable PDF has been overcome by the use of the simulated data from theoretical models. The Integral Equation Model (IEM) [27] was selected because it has the advantage of being applicable to a wide range of roughness scales. The general condition of validity of the model is ks < 3, where k is the wave number (≈ 0.2732 cm^{-1} for 1.3 GHz).

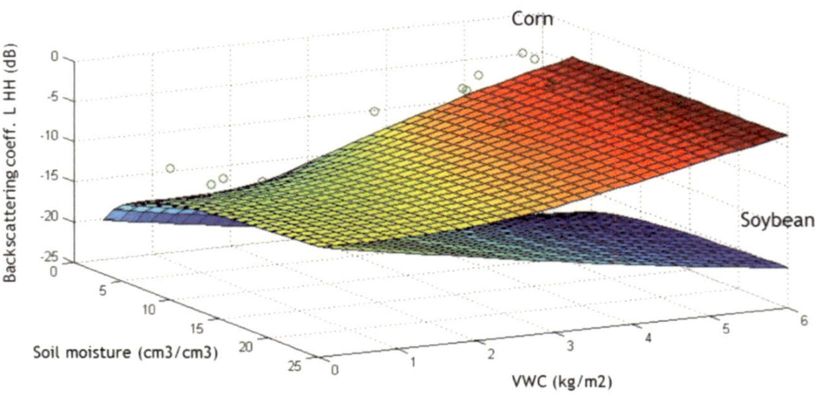

Figure 7. Comparison of measured backscattering coefficients (dots) with simulation from Water Cloud Model (surfaces) after the proper calibration of the free parameters. Two simulated surfaces are reported, one for corn (the red one) and the other soybean (the blue one).

For bare soil, these unknown parameters are the real part of the dielectric constant (ε), the standard deviation of the height (s) and the correlation length (l), the latter two describing the morphology of the surface. For vegetated fields, the Bayesian inversion was performed under two different approaches. In one case the Water Cloud Model (WCM) [35] is used to simulate the backscattering coefficients from vegetation. In the second case, the PDF parameters are properly modified to take into account vegetation effect through empirical relation with vegetation [38]. In both cases, the Vegetation Water Content (VWC) is added as unknown parameter, and it is derived from optical images. In this way, the approach exploits a synergy between SAR and optical images.

At the beginning, the conditional probability is assumed as normal distribution. In the training phase, the conditional PDF is evaluated using measured data (f_{im}) and simulated values from the IEM model (f_{ith}). The distribution assumption is then verified with a chi-squared statistics. The noise function N_I (eq. 4) and the related PDF parameters (mean and standard deviation) are calculated from the statistics of the ration between measured and simulated backscattering coefficients as follows [11, 39]:

$$N_i = \frac{f_{im}}{f_{ith}}. \tag{4}$$

Subsequently a joint PDF is obtained as a convolution of single independent PDFs. The joint PDF is a posterior probability derived from prior probability on roughness and soil moisture values and to the conditional probability which relates the variations of backscattering coefficients to variations of soil moisture and roughness. The relationship can be expressed as follows:

$$p\left(S_i \middle| \sigma_i^0\right) = \frac{p_{prior}\left(S_i\right) p_{post}\left(\sigma_i^0 \middle| S_i\right)}{\int_{S_i} p_{prior}\left(S_i\right) p_{post}\left(\sigma_i^0 \middle| S_i\right) dS_i}, \tag{5}$$

where the term at the denominator is a normalization factor with integration over all variables $S_{i.}$ The variables Si can be:

- For bare soil: dielectric constant (ε), the standard deviation of the height (s) and the correlation length (l);
- For vegetated soil: dielectric constant (ε), the standard deviation of the height (s) and the correlation length (l), vegetation water content (*VWC*).

The variables σ^0_i refer to the input values derived from remote sensing data, which in the presented approach are:

- Backscattering coefficients at L-band HH and VV pol for the Argentinean test sites;

- Backscattering coefficients at C-band HH and VV pol, L-band HH and VV pol and the combination of C and L band at HH pol for SMEX'02 test site.

Based on the field data, the integration ranges for Bayesian inference were selected with different values as is illustrated in the following part. The main aim of using different intervals was to test the sensitivity of the methods to prior information, Through these integrations, to each pixel a value of dielectric constant is associated, starting from the corresponding back-scattering coefficient values. Finally, with the formula proposed by [40] the dielectric constant values have been transformed to estimated values of soil moisture. The flowchart in Fig.8 outlines the main steps of the algorithm, including training and test phase.

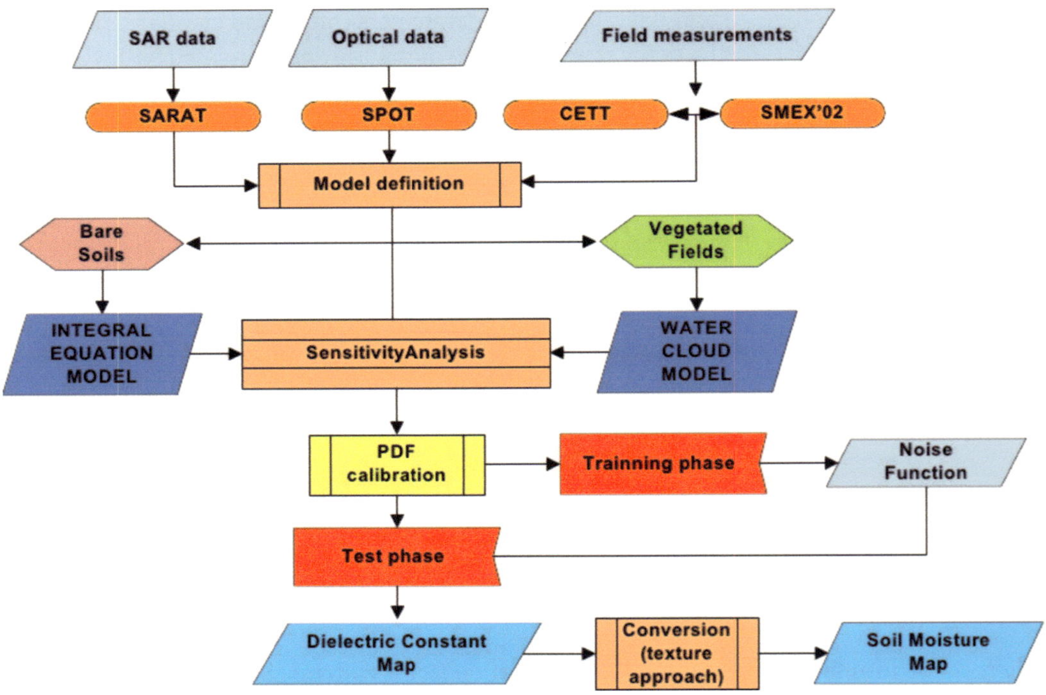

Figure 8. Flowchart of the Bayesian soil moisture approach applied to the Argentinean test site.

As above mentioned, another version of the Bayesian algorithm was developed to take into account the effect of vegetation into the PDF. The flowchart of the algorithm is the same as shown in fig.8, but instead of Water Cloud Model there is an adaptation of the PDF mean to an empirical function related to VWC as detailed described in Notarnicola et al., 2007 [38]. The algorithm was developed to work with C, L and combination of C and L band. In this work, this specific algorithm is applied to SMEX'02 data.

4. Results and discussion

The main aim of the work is to verify the sensitivity of the algorithm to prior conditions of roughness and vegetation in order to optimize the accuracy of the results. Based on this concept several retrievals were performed for different conditions of surface roughness, with specific algorithms for each coverage type in the study area. In the following the results on the Argentinean and SMEX'02 test sites are presented and discussed.

4.1. Argentinean test site

Over the Argentinean test site, the algorithm (fig.8) is divided in two main parts: one to be used in plots with bare soil or covered with sparse vegetation and another for vegetated soils. In both cases, two versions of the algorithm were developed: a simplified one working on a vector of mean values for each plot where the aim is to analyze the backscatter coefficient behavior using random values within ranges of s and l, and another one to work on the whole image, on pixel basis, to investigate the SM spatial distributions. Working with average values of backscattering coefficients has two objectives: to understand the effect on the SM estimates when the signal noise in the single plot is strongly reduced and to lower the computation burden when applying a random function for s and l variables.

An extensive analysis was conducted in order to understand the behavior of variables such as surface roughness and vegetation presence in the final SM estimation through the variability of the prior information. The different cases analyzed are listed below:

- Case 1: Pixel based algorithm for bare soil with fixed roughness. Three runs were executed: s=0.3 cm, l=5.0 cm; s=0.5 cm, l=5.0 cm and s=0.9 cm, l=5.0 cm. Then a mean value map is generated.
- Case 2: Pixel based algorithm for bare soil with an integration over a roughness range: 0.6 cm < s < 1.4 cm; l=5.0 cm.
- Case 3: Pixel based algorithm for bare soil with an integration over a roughness range: 0.6 cm < s < 1.4 cm; l=15.0 cm.
- Case 4: Pixel based algorithm for bare soil with an integration over a roughness range: 1.0 cm< s < 1.5 cm; l=5.0 cm. In this case a very small integration range was considered.
- Case 5: Algorithm applied to backscattering coefficients averaged at plot level with a random function. Values range: 0.5 cm < s < 1.2 cm; 5.0 cm < l < 10.0 cm.
- Case 6: VWC is calculated using a SPOT image. Fixed roughness and correlation length. s=0.5 cm; l=5.0 cm.
- Case 7: VWC is calculated using a SPOT image and a random function is implemented for s and l calculation, considering expected mean and standard deviation values for each parameter: mean value of s=0.7 cm and standard deviation value of 0.5 cm, mean value of l=5.0 cm and standard deviation of 5.0 cm. The random function is built as a noise function

added to the mean values of *s* and *l*. The pseudo random values are drawn from a standard normal distribution.

- Case 8: VWC is provided as an input variable and an integration is done over the following values: 0.01 <VWC < 6 Kg m^2.
- Case 9: VWC is calculated using a SPOT image, based on NDVI values, and an integration is done over roughness and correlation length in the following ranges:0.4 cm< s < 1.2 cm and 3.0 cm < l < 10.0 cm.

In Fig.9, preliminary results are presented where the different analyzed cases based on various prior conditions are numbered from 1 to 9. In general, for bare soil (fig. 9), the results showed a sensitivity of the algorithms to the different roughness conditions of each plot with a variability of around 5-7% (excluding the extreme cases). The highest variability among the cases is around 40% and is found when the roughness interval is very small (case 2 and 3). When considering a random function for roughness (case 7) and when performing the retrieval over average values of backscattering coefficients (case 5), the mean different with respect to ground measurements is around 15%.

For vegetated areas, due to the limited availability of field measurements (field 5N), the evaluation of the performances is still under work. More extensive results for vegetation are presented for the SMEX'02 experiments.

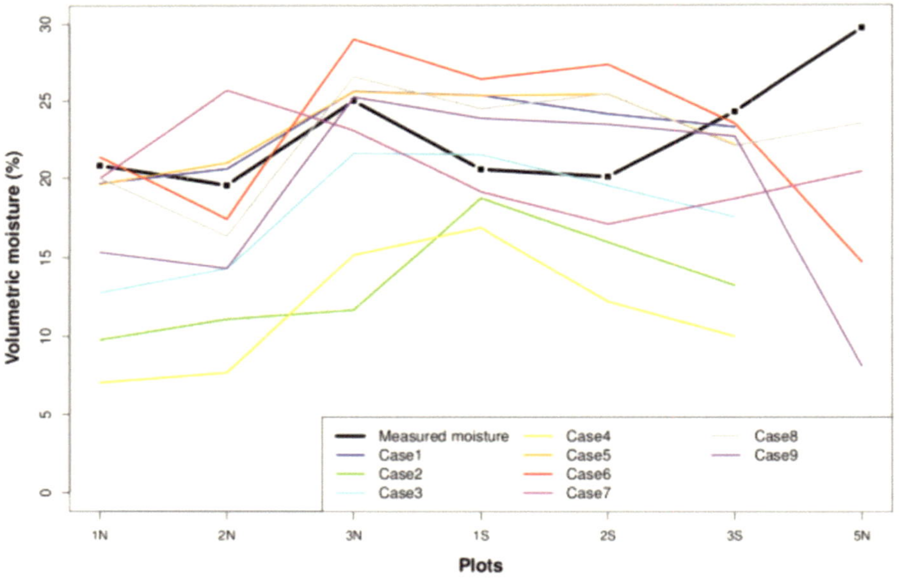

Figure 9. Comparison of SM estimates with measured values. Behavior diagram of the described cases.

http://dx.doi.org/10.5772/57562

Case1 and 3 results are reported in the form of SM maps in fig.10. A detailed analysis of the maps in fig.10 indicated error patterns detected for cases with rows of plots oriented orthogonally to the direction of the sensor observation. As it was observed, the backscattering coefficients for HH polarization is sensitive to the orientation of lines tillage and no inversion algorithms consider this factor. Consequently the results show significant errors in plots perpendicular to the observation.

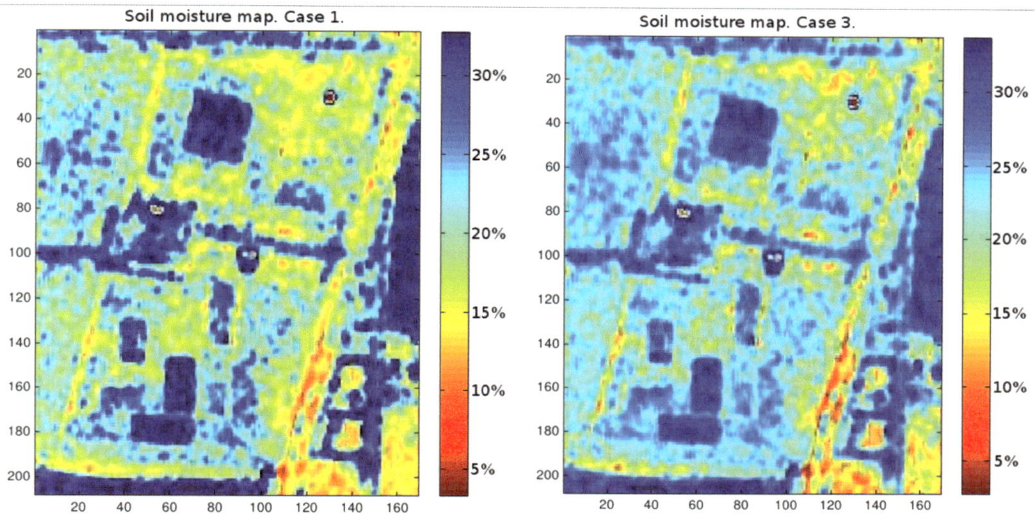

Figure 10. Soil moisture maps for Case 1 (left) and Case 3 (right) over the selected test site.

Case 1 shows that the northern plots with bare soils (1N and 2N) have moisture values very similar to the ground truth. On the contrary, southern plots with bare soils (1S and 2S) have higher moisture values than the measured ones, having the first of them a value of 25%, while the in situ data shown values around 20%. Case 3 shows that plot 1N and 2N obtained moisture values around 15%, which represents an over-estimation of the actual value of around 5%. For plots 1S and 2S, the estimate values are between 22 to 24%. Case 1 could model with good accuracy plots 1N and 2N losing accuracy in southern plots. On the contrary, Case 3 could model with relatively accuracy plots 1S and 2S losing precision in northern plots. The factor of apparent roughness change can be attributed to the orientation of the rows with respect to the SAR signal [41].

4.2. Results on the SMEX'02 experiments

As illustrated in previous paragraphs, the inversion methodologies based on Bayesian approach can be applied to different sensors configurations. In this way different polarizations and/or bands can be exploited to extract soil features. In fact, due to the different way C band or L band signals interact with soil and the above canopy layer, they are sensitive to different surface characteristics. Thus a proper combination of the two bands can help disentangle the effect of vegetation and then improve the estimation of soil moisture. In this paragraph, the results of the Bayesian methodologies are illustrated and the evaluated in terms of:

- Correlation coefficients, R, between the estimates and the ground truth values
- Root Mean Square Error, RMSE, between the estimates and the ground truth values.

When dealing with the different cases due to the prior information, the retrieved values will be compared with the measurements through the Taylor plots [42].

The Bayesian approach has been applied to AirSAR data collected during the SMEX'02 experiments considering C band, L band and combination of C and L data.

The results for the estimation of SM are reported in table 4. As expected the estimation of SM is quite difficult, thus determining values of R varying from 0.47 to 0.80 for the combination of C and L bands. The highest difficulties are found for the detection and correct estimation of extreme values of soil moisture.

Configurations	**Correlation coefficients**	**RMSE (cm^3/cm^3)**
C band	0.47	0.10
L band	0.67	0.05
C + L band	0.80	0.02

Table 4. Correlation coefficients (R), RMSE for the comparison between the different estimates and ground truth values for SM values, excluding extreme values.

The retrieval of low values of SM can be difficult as the signal for soil is small and difficult to be disentangled from the vegetation signal. For high values, the signal from soil is strong but in the case of C bands the double bouncing and the effect of absorption from leaves also for L band, typical of narrow leaf plants such as soybean, determine a lower signal reaching the sensor [43]. The L band estimates are the only one able to predict highest values of SM.

Similar analyses were also found in Notarnicola et al. 2006 [39]. In this previous analysis, the methodologies were applied only to few fields of the same data set. With respect to the accuracy reported in Notarnicola et al., 2006 [39], a worsening in the performance is found. In particular the data set includes all the fields in the watershed basin and the fields located in the eastern part which exhibits anomalous values of SM, some very high values around 35% and some values lower than 5%. Considering the available meteorological information, the eastern and western parts of the watershed experienced very different intensity for the rain event where most of the rain event occurred in the western part.

If the watershed is divided in two parts the western and the eastern part the performances of the algorithm for SM retrieval differ significantly. The correlation coefficients are equal to 0.57 and 0.84, not significantly different from those found in [39].

Furthermore, the performances notably change if in the data set the soybean and corn fields are considered separately. The results are reported in table 5.

Configurations	Corn		Soybean	
	R	**RMSE (cm^3/cm^3)**	**R**	**RMSE (cm^3/cm^3)**
C band	0.36	0.127	*0.83*	*0.032*
L band	0.41	0.091	0.42	0.072
C+L band	*0.68*	*0.057*	*0.82*	*0.037*

Table 5. Correlation coefficients (R), RMSE for the comparison between the different estimates and ground truth values for SM values and for the Bayesian approach. With respect to table 4, in this case, the soybean and corn fields are considered separately. In italics, the values significantly different from the one found in whole data sets are indicated.

Similar characteristics are also found in [44], where it is proved that the RMSE is dependent on the level of vegetation of the different fields. Furthermore, in the case of C band, the signal coming from the VWC dominates over the signal coming from soil. In fact, when the vegetation has low value of VWC such as in the case of soybean fields, the C band is able to provide acceptable estimates for soil moisture. In the case of corn fields, the best results is obtained with the combination of C and L band, one sensitive to VWC and the other to the surface contribution. These discrepancies may be ascribed to the fact that in the Bayesian formulation the double bouncing between soil and corn trunk effect is not taken into account. This effect in such kind of plants with broad leaves could dominate [43].

On the SM estimates derived from combination of C and L band, a further analysis has been carried out by considering the effect of prior information on roughness.

More in details, the range of roughness in the integration of equation which is used to derive the expected values for soil moisture has been varied as follows:

- Low roughness: s varying between 0.2 and 1.2 cm;
- High roughness: s varying between 1.2 and 5.0 cm;
- Whole range of roughness: s varying between 0.2 and 5.0 cm.

The chosen values for roughness have selected based on prior information on roughness during field measurements. Along with these fixed ranges of roughness, a variable roughness interval has been considered based on the values of backscattering roughness. Higher values of backscattering coefficients on both C and L band have been also associated to high values of roughness.

The SM estimates derived from C and L band are illustrated in figure 11. When the estimates under these hypotheses are compared, they show an overall variability of around 25%. The results in term of correlation coefficients, are presented in the form of Taylor diagram as showed in fig. 12.

The SM estimates closest due to the ground measurements are those derived from the whole range of roughness and the adapted intervals. The high roughness and the whole range of roughness produces very close results both in terms of correlation coefficients and standard deviations.

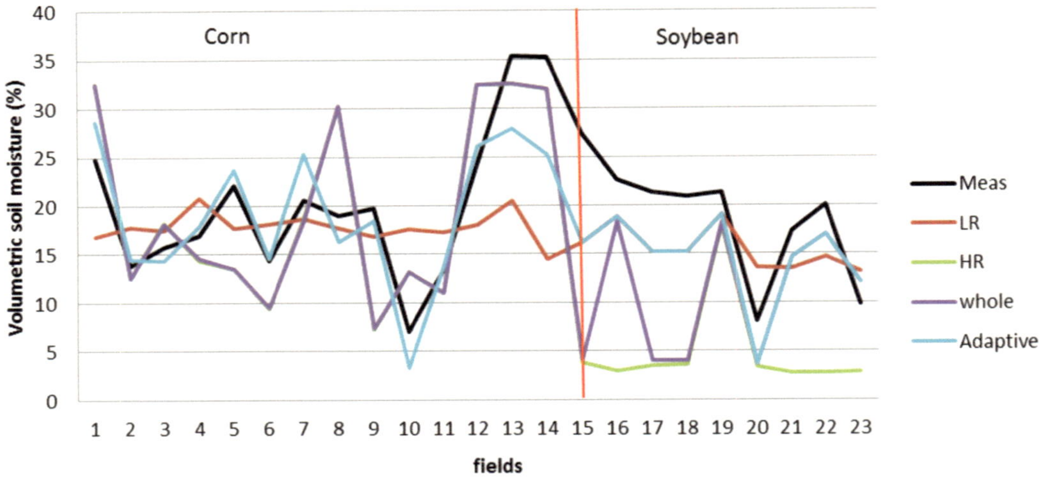

Figure 11. Comparison of SM estimates under different roughness hypothesis with ground measurements. "LR" stand for low roughness, "HR" for high roughness, "Whole" for the whole range of roughness and adaptive for adaptive values of roughness.

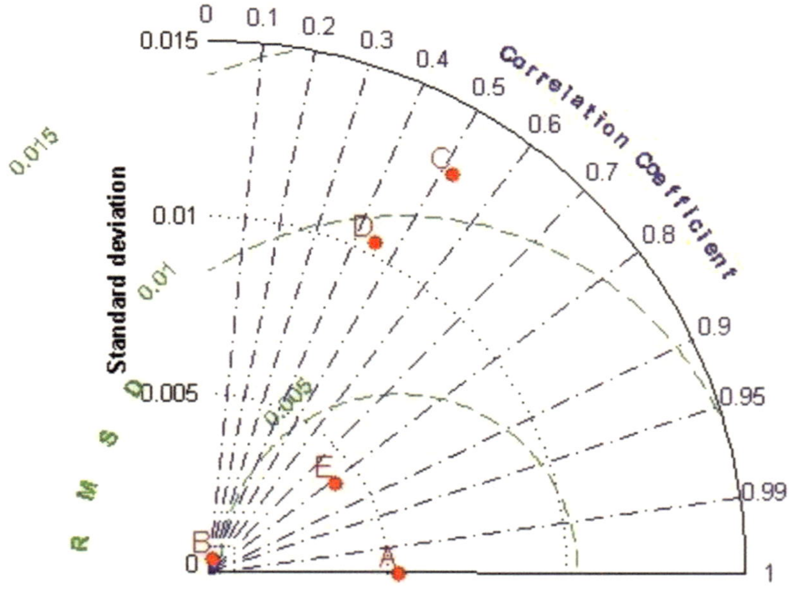

Figure 12. Taylor diagram showing the comparison under different prior hypotheses on roughness. A refers to ground measurements; B to low roughness; C to high roughness; D to whole range of roughness; E adapted roughness ranges.

http://dx.doi.org/10.5772/57562

5. Conclusions

The main objective of this chapter is to present the capability of Bayesian approach to estimate SM values starting from SAR backscattering coefficients. Two case studies are presented where SAR acquisitions took place over agricultural fields. The first case study was related to an Argentinean test site developed and equipped for acquisitions of airborne L band SAR called SARAT. The acquisitions were carried out in preparation of the SAOCOM mission. The second case study was related to the experiment SMEX'02 carried out in IOWA in 2002. In this experiment airborne AirSAR images were available and for this reason the retrieval was applied to C band, L band and C+L band data.

Based on the retrieval results, the main goal was then to verify the sensitivity of the SM estimates from the set prior information on roughness and vegetation. All the prior PDFs are set a uniform, non-informative but the set limits of the interval in the integration procedure can determine variation in the final SM estimates. This behavior is expected because the electromagnetic models used in the retrieval approach contain explicitly the dependence of backscattering coefficients on roughness and vegetation parameters.

The effect of prior information ranges from few percentages up to 25% where the highest sensitivity is found in both case studies when too specific and narrow intervals for roughness are used. The highest performances were found for both case studies when the range of roughness is large enough to include most roughness measurements. Moreover, if a preliminary assessment on the roughness level is available, the algorithm determines the highest performances with respect to the ground reference data.

An interesting feature observed in the case of Argentinean test site is the reduction of errors on the SM estimates when the retrieval is performed on average values of backscattering coefficients from each field. This behavior can be due to the reduction of noise present in the SAR signal.

As main conclusion of this analysis and suggestions in using the proposed the Bayesian algorithms for SM estimation, the following considerations emerge:

- The set of prior information has to be selected carefully;
- Even in the case of non-information prior PDF, the range of variability of the prior variable has an impact on the final estimates;
- It is preferable to integrate over a large interval of roughness and/or vegetation variables in order to take into account and properly weight all the measured values.
- As the speckle noise can influence the SM estimates, a proper filter over the SAR image needs to be applied before proceeding to the retrieval approach.

Acknowledgements

The authors wish to thank to CONAE and SAOCOM mission team for SARAT data provision and field measurements acquired during the remote sensing campaign.

Author details

Claudia Notarnicola[1*] and Romina Solorza[2]

*Address all correspondence to: claudia.notarnicola@eurac.edu

1 EURAC Research-Institute for Applied Remote Sensing, Viale Druso, Bolzano, Italy

2 Gulich Institute, CONAE, Falda del Carmen, Córdoba, Argentina

References

[1] Barrett B.W., Dwyer E., Whelan P. Soil Moisture Retrieval from Active Spaceborne Microwave Observations: An Evaluation of Current Techniques. Remote Sens., 2009; 1 210-242, doi:10.3390/rs1030210.

[2] World Meteorological Observatory, Climate Essential Variable http://www.wmo.int/pages/prog/gcos/index.php?name=EssentialClimateVariables (access 5 November 2013).

[3] Álvarez Mozos J., Casalí J., González A., López J. Estimación de la Humedad Superficial del Suelo mediante Teledetección de Radar en Presencia de una Cubierta de Cereal. Estudios de la Zona No Saturada del Suelo, 2005; VII 313–318.

[4] Engman E.T. Applications of microwave remote sensing of soil moisture for water resources and agriculture. Remote Sensing of Environment, 1991; 35 213–226,.

[5] Ulaby F.T., Aslam A., Dobson M.C. Effects of vegetation cover on the radar sensivity to soil moisture. IEEE Transactions on Geoscience and Remote Sensing, 1982; 20(4) 476–481.

[6] Lakshmi V. Remote Sensing of Soil Moisture. Hindawi Publishing Corporation ISRN Soil Science Volume 2013, Article ID 424178, 33 pages.

[7] Gruhier C., de Rosnay P., Hasenauer S., Holmes T., de Jeu R., Kerr Y., Mougin E., Njoku E., Timouk F., Wagner W., Zribi M. Soil moisture active and passive microwave products: intercomparison and evaluation over a Sahelian site, Hydrol. Earth Syst. Sci., 2010; 14 141–156.

[8] Behari J. Microwave Dielectric Behavior of Wet Soils. Anamaya Publishers, 2005.

[9] Ulaby F. T. Radar measurement of soil moisture content. IEEE Transactions on Antennas and Propagation, 1974; 22(2) 257–265. doi: 10.1109/TAP.1974.1140761.

[10] Ulaby F.T., Moore R.K., Fung A.K. Microwaves Remote Sensing: Active and Passive., volume III, Surface Scattering and Emission Theory. Artech House, Dedham, MA, 1986.

[11] Dubois P., Van Zyl J., Engman T. Measuring soil moisture with imaging radars. IEEE Transactions on Geoscience and Remote Sensing, 1995; 33 (4) 915–926.

[12] Baghdadi N., Aubert M., Zribi M. Use of TerraSAR-X Data to Retrieve Soil Moisture Over Bare Soil Agricultural Fields. IEEE Geoscience and Remote Sensing Letters, 2012; 9(3) 512–516.

[13] Dobson F.T. Ulaby L.E. Pierce, T.L. Sharik K.M. Bergen J. Kellindorfer J.R. Kendra E. Li Y.C. Lin A. Sarabandi K., Siqueira P.. Estimation of forest byophysical characteristics in northern michigancwith sir-c/x-sar. IEEE Transactions on Geoscience and Remote Sensing, 1995; 33 877–895.

[14] Ferrazzoli P., Paloscia S., Pampaloni P., Schiavon G. The potential of multifrequency polarimetric sar in assesing agricultural and arboreous biomass. IEEE Transactions on Geoscience and Remote Sensing, 1997; 35 5–17.

[15] Baronti S., Del Frate F., Ferrazzoli P., Paloscia S., Pampaloni P., Schiavon G.. SAR polarimetric features of agricultural areas. International Journal of Remote Sensing, 1995; 16 2639–2356.

[16] Holah N., Baghdadi N., Zribi M., Bruand A., King C.. Potential of ASAR/Envisat for the characterization of soil surface parameters over bare agricultural fields. Remote Sensing of Environment, 2005; 96 78–86.

[17] Li Z., Ren X., Li X., Wang L. Soil moisture measurement and retrieval using Envisat ASAR imagery. Proceedings of the IEEE International Geoscience and Remote Sensing Symposium (IGARSS 2004), 3539–3542, Anchorage, USA, 2004.

[18] Zhang T., Zeng Q., Li Y., Xiang Y. Study on relation between InSAR coherence and soil moisture. Proceedings of the ISPRS Congress, Beijing, China, 2008.

[19] Autret, Bernard R., VidalMadjar D.. Theoretical study of the sensitivity of the microwave backscattering coefficient to the soil surface parameters. International Journal of Remote Sensing, 1989; 10 171–179.

[20] Chen K.S., Yen S.K., Huang W.P.. A simple model for retrieving bare soil moisture from radar scattering coefficients. Remote Sensing of Environment, 1995; 54 121–126.

[21] Pierdicca N., Pulvirenti L., Bignami C.. Soil moisture estimation over vegetated terrains using multitemporal remote sensing data. Remote Sensing of Environment, 2010; 114 440–448.

[22] Balenzano A, Satalino G., Belmonte A., D'Urso G., Capodici F., Iacobellis V., Gioia A., Rinaldi M., Ruggieri S., Mattia F. On the use of multitemporal series of COSMO-SkyMed data for LANDcover classification and surface parameter retrieval over agricultural sites. Proceeding of Geoscience and Remote Sensing Symposium (IGARSS), 142–145, July 2011.

[23] Hornacek M, Wagner W, Sabel D, Truong H.L, Snoeij P, Hahmann T, Diedrich E, Doubkova M. Potential for High Resolution Systematic Global Surface Soil Moisture Retrieval Via Change Detection Using Sentinel-1. IEEE Journal of Selected Topics in Applied Earth Observations and Remote Sensing, 2012, 1–9.

[24] Dobson M.C., Ulaby F. T.. Mapping Soil Moisture Distribution With Imaging Radar, chapter 8, 407–433. John Wiley and Sons Inc., 1998.

[25] Shi J., Wang J., Hsu A.Y., ONeill P.E., Engman E.T.. Estimation of bare surface soil moisture and surface roughness parameter using l-band SAR image data. IEEE Transactions on Geoscience and Remote Sensing, 1997; 35 1254–1266.

[26] Oh Y.. Quantitative retrieval of soil moisture content and surface roughness from multipolarized radar observations of bare soils surfaces. IEEE Transactions on Geoscience and Remote Sensing, 2004; 42(3) 596–601.

[27] Fung, A. K. Microwave Scattering and Emission Models and their Application. Artech House, Boston; 1994.

[28] Pampaloni P., Santi E., Paloscia S., Pettinato S., Poggi P. Radar Remote Sensing of Soil Moisture, ENVISNOW Project, SMC algorithms. Technical report, IFAC-CNR, October 2004.

[29] Notarnicola C., Posa F. Bayesian algorithm for the estimation of the dielectric constant from active and passive remotely sensed data. IEEE Geoscience and Remote Sensing Letters, 2004; 1(3) 179–183.

[30] Paloscia S., Pampaloni P., Pettinato S., Santi E. A comparison of algorithms for retrieving soil moisture from ENVISAT/ASAR images. IEEE Transactions on Geoscience and Remote Sensing. 2008; 46(10) 3274-3284.

[31] Barber M., Grings F., Perna P., Bruscantini C., Karszenbaum H. Análisis de las mediciones de rugosidad del Centro Espacial Teófilo Tabanera, Córdoba. Technical report, CONAE-IAFE,Mayo

[32] http://nsidc.org/data/amsr_validation/soil_moisture/smex02/

[33] Barber M., Grings F., Perna P., Piscitelli M., Maas M., Bruscantini C., Jacobo-Berlles J., Karszenbaum H. Speckle Noise and Soil Heterogeneities as Error Sources in a Bayesian Soil Moisture Retrieval Scheme for SAR Data. IEEE Journal Of Selected Topics In Applied Earth Observations And Remote Sensing, 2012; 5(3).

[34] Gelman A., Carlin J.B., Stern H.S., Rubin D. B., Bayesian data analysis, Chapman&Hall/CRC 1995.

[35] Attema E. P, Ulaby F.T.. Vegetation Modeled as a Water Cloud. Radio Science, 1978; 13(2) 357–364.

[36] Dabrowska-Zielinska K., Inoue Y., Kowalik W., Gruszczynska M.. Inferring the effect of plant and soil variables on C-and L-band SAR backscatter over agricultural fields, based on model analysis. Advances in Space Research, 2007; 39 139–148.

[37] Jackson T., Chen D., Cosh M., Li F., Anderson M., Walthall C., Doriaswamy P., Ray Hunt E. Vegetation water content mapping using Landsat data derived normalized difference water index for corn and soybeans. Remote Sensing of Environment, 2004; 92 475–482.

[38] Notarnicola, C. & Posa, F. Inferring vegetation water content from C and L band images, IEEE Transactions on Geoscience and Remote Sensing, 2007; 45(10) 3165-3171.

[39] Notarnicola C., Angiulli M., Posa F. Use of Radar and Optical Remotely Sensed Data for Soil Moisture Retrieval on Vegetated Areas. IEEE Transactions on Geoscience and Remote Sensing, 2006; 44(4) 925-935.

[40] Hallikainen M.T., Ulaby F.T., Dobson M.C., El Rayes M.A., Wu L.K.. Microwave dielectric behavior of wet soil-part 1: Empirical models and experimental observations. IEEE Transactions On Geoscience And Remote Sensing, 1985; GE-23(1) 25–34.

[41] Solorza R., Notarnicola C., Karszenbaum H., Retrieval of soil moisture using a Bayesian approach and electromagnetic models in views of the SAOCOM mission: study on SARAT images in an agricultural site in Argentina, Proceeding of Igarss 2013, July 2013, Melbourne, Australia.

[42] Taylor, E, K, Summarizing multiple aspects of model performance in a single diagram, Journal of Geophysical Research, 2001; 106(D7), 7183-7192.

[43] Macelloni G., Paloscia S., Pampaloni P., Marliani F., Gai, M. The relationship between the backscattering coefficient and the biomass of narrow and broad leaf crops, IEEE Transaction on Geoscience and Remote Sensing, 2001; 39(4) 873-884.

[44] Lakhankar T., Ghedira H., Temimi M., Sengupta M., Khanbilvar R., Blake, R. Nonparametric methods for soil moisture retrieval from satellite remote sensing data, Remote Sensing, 2009; 1, 3-21. doi: 10.3390/rs1010003.

Section 2

Applications of Dynamic Programming

Chapter 3

Optimizing Basel III Liquidity Coverage Ratios

J. Mukuddem-Petersen, M.A. Petersen and
M.P. Mulaudzi

Additional information is available at the end of the chapter

http://dx.doi.org/10.5772/58395

1. Introduction

The recent financial crises, viz., the subprime mortgage and 2007-2009 financial crises as well as the sovereign debt crisis, are characterized by too-big-to fail banks that suffered from a lack of liquidity (see, for instance, [13] and [14]). The actualization of such liquidity risk led to credit crunches and had negative effects on global financial markets. In response to this, among other things, the Basel Committee on Banking Supervision (BCBS) is proposing that banks should always have a 30-day liquidity cover for stress scenarios (see, for instance, [2], [3], [4] and [5]).

In this regard, the level of high-quality liquid assets (HQLAs) is important in order for banks to function optimally (see, for instance, [6] and [8]). As far as Basel III liquidity proposals are concerned, the BCBS is suggesting a liquidity coverage ratio (LCR) defined as follows.

The LCR has two components:

(a) total stock of HQLAs; and

(b) total nett cash outflows,

and is expressed as

$$\text{LCR} = \frac{\text{Total Stock of High-Quality Liquid Assets (HQLAs)}}{\text{Total Nett Cash Outflows (NCOs) Over the Next 30 Calendar Days}} \geq 1. \qquad (1)$$

The numerator of the LCR is the stock of HQLA. Under the standard, banks must hold a stock of unencumbered HQLA to cover the total net cash outflows over a 30-day period under the prescribed stress scenario. In order to qualify as HQLA, assets should be liquid in markets during a time of stress and, in most cases, be eligible for use in central bank operations. Certain types of assets within HQLA are subject to a range of haircuts. HQLA are comprised of Level 1 assets (L1As) and Level 2 assets (L2As). L1As generally include cash, central bank reserves, and certain marketable securities backed by sovereigns and central banks, among others. These assets are typically of the highest quality and the most liquid, and there is no limit on the extent to which a bank can hold these assets to meet the LCR. L2As are comprised of Level 2A assets (L2AAs) and Level 2B assets (L2BAs). L2AAs include, for example, certain government securities, covered bonds and corporate debt securities. L2BAs include lower rated corporate bonds, residential mortgage backed securities and equities that meet certain conditions. L2As may not in aggregate account for more than 40 % of a bank's stock of HQLA. L2BAs may not account for more than 15 % of a bank's total stock of HQLA (see, for instance, [1] and [13]).

The denominator of the LCR is the total net cash outflows. It is defined as total expected cash outflows, minus total expected cash inflows, in the specified stress scenario for the subsequent 30 calendar days. Total expected cash outflows are calculated by multiplying the outstanding balances of various categories or types of liabilities and off-balance sheet commitments by the rates at which they are expected to run off or be drawn down. Total expected cash inflows are calculated by multiplying the outstanding balances of various categories of contractual receivables by the rates at which they are expected to flow in. Total cash inflows are subject to an aggregate cap of 75 % of total expected cash outflows, thereby ensuring a minimum level of HQLA holdings at all times (see, for instance, [11] and [13]).

The standard requires that, in the absence of financial stress, the value of the ratio be no lower than 100 % (i.e., the stock of HQLA should at least equal total net cash outflows). Banks are expected to meet this requirement on an ongoing basis and hold a stock of unencumbered HQLA as a defence against the potential onset of liquidity stress. During a period of financial stress, however, banks may use their stock of HQLA, thereby falling below 100 % (see, for instance, [11]).

1.1. Overview of the literature

Our contribution has strong connections with [7], [10], [12] and [13] that deals with subprime mortgage funding and liquidity risk and Basel III liquidity regulation, respectively.

The working paper [7] examines large capital injections by U.S. financial institutions from 2000 to 2009. These infusions include private as well as government cash injections under the Troubled Asset Relief Program (TARP). The sample period covers both business cycle expansions and contractions, and the recent financial crisis. Elyasiani, Mester and Pagano show that more financially constrained institutions were more likely to have raised capital through private market offerings during the period prior to TARP, and firms receiving a TARP injection tended to be riskier and more levered. In the case of TARP recipients, they appeared to finance an increase in lending (as a share of assets) with more stable financing sources such as core deposits, which lowered their liquidity risk. However, in [7], Elyasiani, Mester and Pagano find no evidence that banks' capital adequacy increased after the capital injections. In this book chapter, we regard the LCR as a measure of liquidity risk. We check the tendencies of this measure over the sample period 2002 to 2012 and make conclusions about it (see, also, [13]).

The technique employed in this book chapter is heavily reliant on the one used in [10]. In that paper, Mukuddem-Petersen and Petersen consider the application of stochastic optimization theory to asset and capital adequacy management in banking. The study is motivated by new banking regulation that emphasizes risk minimization practices associated with assets and regulatory capital. The analysis in [10] depends on the dynamics of the capital adequacy ratio (CAR), that we compute by dividing regulatory bank capital (RBC) by risk weighted assets (RWAs). Furthermore, Mukuddem-Petersen and Petersen demonstrate how the CAR can be optimized in terms of bank equity allocation and the rate at which additional debt and equity is raised. In either case, the dynamic programming algorithm for stochastic optimization is employed to verify the results. Also, in [10], Mukuddem-Petersen and Petersen provide an illustration of aspects of bank management practice in relation to this regulation. In the current chapter, the same technique is employed (see, also, [13]).

In [12], we use actuarial methods to solve a nonlinear stochastic optimal liquidity risk management problem with deposit inflow rates and marketable securities allocation as controls. The main objective in [12] is to minimize liquidity risk in the form of funding and credit crunch risk in an incomplete market. In order to accomplish this, we construct a stochastic model that incorporates mortgage and deposit reference processes. However, the current chapter is an improvement on [12] in that bank balance sheet features play a more prominent role (see Sections 2, 3 and 5 for more details).

In order to construct our LCR model, we take into account results obtained in [13] in a discrete-time framework (see, also, [8]). In the aforementioned book, we estimate the LCR and NSFR by applying approximation techniques to banking data from a cross section of countries. We find that these Basel III risk measures have low information values and are relatively poor indicators of liquidity risk. Our results, in [13], show that as the LCR increases (decreases) the probability of failure decreases (increases) for both Class I (internationally active banks with Tier 1 capital in excess of US \$ 4 billion) and II (the rest) banks. Our contribution is distinct from the aforementioned in the following respects. Firstly, our analysis has a heavy reliance on the derivation of a stochastic model for LCR dynamics that depends mainly on the liquidity provisioning rate, HQLA returns and NCO outflows. Secondly, we obtain an analytic solution of a particular type to our stochastic bank LCR problem (with a quadratic objective function) that we pose. Finally, the optimal choices for the cash injection and asset allocation are both expressed in terms of a LCR reference process. To our knowledge such processes have not been considered for LCRs before. The study is particularly significant because the Basel III LCR will be implemented on Thursday, 1 January 2015 on a global scale (see, for instance, [2], [3], [4] and [5]). In this book chapter, we extend the analysis in [13] to continuous time.

1.2. Outline of the book chapter

In short, this book chapter advances our knowledge of Basel III liquidity by investigating the LCR global liquidity standard (see, for instance, [6] and [14]) in an optimization context. In particular, in Section 2, a theoretical-quantitative model is constructed by considering the dynamics of the HQLAs and NCOs. Section 3 produces two parameters that are able to be controlled, viz., the liquidity provisioning rate and HQLA allocation. The main motivation for studying LCR dynamics is to show that, in principle, banks are able to control their liquidity via an appropriate provisioning strategy. This should ensure that the said ratio does not move below an acceptable level. The control theoretic liquidity problem is to meet LCR targets with as little additional liquidity provisioning (essentially corresponding to cash injections in our chapter) as possible and optimal HQLA allocation. Section 5 provides conclusions and future directions. As was mentioned before, the additional provisioning may arise from an inflow of cash injections. We choose examples to illustrate that Basel III liquidity regulation resulted from both problematic liquidity structures and unexpected cash outflows (see Section 5 for more details).

2. A Liquidity Coverage Ratio Model

In this section, we model HQLAs, NCOs and LCR in a stochastic framework by following [10] very closely. This is important for solving the optimal LCR control problem outlined subsequently in Section 3.

2.1. Description of the Liquidity Coverage Ratio Model

Before the 2007-2009 financial crisis, banks were prosperous with high liquidity provisioning rates, low interest rates and soaring cash outflows. This was followed by the collapse of the housing market, exploding default rates and the effects thereafter. The LCR was developed to promote short-term resilience of a bank's liquidity risk profile. This standard aims to ensure that a bank has an adequate stock of unencumbered HQLAs that consist of cash or assets that can be converted into cash at little or no loss of value in private markets to meet its liquidity needs for a 30 calendar day liquidity stress scenario (see, for instance, [1], [3] and [4]). In order to make our analysis tractable, we make the following assumption about our LCR model.

Assumption 2.1. (Filtered Probability Space and Time Index) *Assume that we have a filtered probability space* $(\Omega, \mathcal{F}, \mathbf{P})$ *with filtration* $\{\mathcal{F}_t\}_{t\geq 0}$ *on a time index set* $T = [t_0, t_1]$.

Subsequently, we study a system of stochastic differential equations (SDEs) that value HQLAs at time t as $x^1 : \Omega \times T \to \mathbb{R}^+$ (compare with [10]). Here, HQLAs, x_t^1, are stochastic because they are dependent on the stochastic rates of return on L1As and L2As (see [12] for more details). Also, NCOs at time t, x_t^2, with $x^2 : \Omega \times T \to \mathbb{R}^+$ are stochastic because their value has a reliance on random cash in- and outflows as well as liquidity provisioning. Furthermore, for $x : \Omega \times T \to \mathbb{R}^2$ we use the notation x_t to denote

$$x_t = \begin{bmatrix} x_t^1 \\ x_t^2 \end{bmatrix}$$

and present the LCR, $l : \Omega \times T \to \mathbb{R}^+$, by

$$l_t = x_t^1 / x_t^2 = x_t^1 .(x_t^2)^{-1}. \quad (2)$$

It is important for banks that l_t in (2) has to be sufficiently high to ensure high LCRs. In fact, as was mentioned before, Basel III sets the minimum value of the LCR at 1. Obviously, low values of l_t indicate that the bank has decreased liquidity and is at high risk of causing a credit crunch (see, for instance, [13]).

Bank liquidity has a heavy reliance on liquidity provisioning rates. This rate should be reduced for high LCRs and increased beyond the normal rate when bank LCRs are low. In the sequel, the stochastic process $u^1 : \Omega \times T \to \mathbb{R}^+$ is the *normal rate of liquidity provisioning per monetary unit of the bank's NCOs* whose value at time t is denoted by u^1_t. In this case, $u^1_t dt$ is the normal liquidity provisioning rate per unit of the bank's NCOs over the time period $(t, t+dt)$. A related concept is the *adjustment to the rate of liquidity provisioning per monetary unit of the bank's NCOs for surplus or deficit*, $u^2 : \Omega \times T \to \mathbb{R}^+$, that depends on the LCR. In the case of liquidity deficit, during stress scenarios, this adjustment rate can correspond to a cash injection rate. Here the amount of surplus or deficit is reliant on the excess of HQLAs over NCOs. We denote the sum of u^1 and u^2 by the *liquidity provisioning rate* $u^3 : \Omega \times T \to \mathbb{R}^+$, i.e.,

$$u^3_t = u^1_t + u^2_t, \text{ for all } t. \tag{3}$$

The following assumption is made in order to model the LCR in a stochastic framework (compare with [10]).

Assumption 2.2. (Liquidity Provisioning Rate) *The liquidity provisioning rate, u^3, is predictable with respect to $\{\mathcal{F}_t\}_{t\geq 0}$ and provides us with a means of controlling bank LCR dynamics (see (3) for more details).*

The closed loop system will be defined such that Assumption 2.2 is met, as we shall see in the sequel. In times of deficit, for (3), we should choose the cash injection rate, u^2, sufficiently large in order to guarantee bank liquidity. In reality, cash injections are subject to more stringent conditions (see, also, [13]).

Before and during the financial crisis, the LCR decreased significantly as extensive cash outflows took place with a consequent rising of NCOs. By contrast, banks predicted continued growth in the financial markets (see, for instance, [14]). The dynamics of the *outflows per monetary unit of the bank's NCOs*, $e : \Omega \times T \to \mathbb{R}$, is given by

$$de_t = r^e_t dt + \sigma^e_t dW^e_t, \quad e(t_0) = e_0, \tag{4}$$

where e_t is the outflows per NCO monetary unit, $r^e : T \to \mathbb{R}$ is the rate of outflows per monetary unit of the bank's NCOs, the scalar $\sigma^e : T \to \mathbb{R}$, is the volatility in the outflows per NCO unit and $W^e : \Omega \times T \to \mathbb{R}$ is standard Brownian motion (compare with [10]). Moreover, we consider

$$dh_t = r^h_t dt + \sigma^h_t dW^h_t, \quad h(t_0) = h_0, \tag{5}$$

where the stochastic processes $h : \Omega \times T \to \mathbb{R}^+$ is the *investment return on bank HQLAs per monetary unit of HQLAs*, $r^h \to \mathbb{R}^+$ is the rate of HQLA return per HQLA unit, the scalar $\sigma^h : T \to \mathbb{R}$, is the volatility in the rate of HQLA returns and $W^h : \Omega \times T \to \mathbb{R}$ is standard Brownian motion. Before the 2007-2009 financial crisis, riskier HQLA returns were much higher than those of riskless reserves, making the former a more attractive but much riskier investment (see, also, [13]). During and after the crisis, this tendency reversed.

Assumption 2.3. (HQLA Classes) *Suppose from the outset that bank HQLAs can be classified into $n+1$ asset classes. One of these HQLAs is risk free (like Central Bank reserves) while the HQLAs $1, 2, \ldots, n$ have some risk associated with them.*

Riskier HQLAs evolve continuously in time and are modeled using a n-dimensional Brownian motion. In this multidimensional context, the *investment returns on bank HQLAs in the k-th HQLA per monetary unit of the k-th HQLA* is denoted by y_t^k, $k \in \mathbb{N}_n = \{0, 1, 2, \ldots, n\}$ where $y : \Omega \times T \to \mathbb{R}^{n+1}$. Thus, the return per HQLA unit may be given by

$$y = (\mathtt{R}(t),\ y_t^1,\ \ldots,\ y_t^n),$$

where $\mathtt{R}(t)$ represents the return on reserves and $y_t^1, \ \ldots, \ y_t^n$ represent riskier HQLA returns. Furthermore, we can model y as

$$dy_t = r_t^y dt + \Sigma_t^y dW_t^y, \ \ y(t_0) = y_0, \tag{6}$$

where $r^y : T \to \mathbb{R}^{n+1}$ denotes the rate of asset returns, $\Sigma_t^y \in \mathbb{R}^{(n+1)\times n}$ is a covariance matrix of HQLA returns and $W^y : \Omega \times T \to \mathbb{R}^n$ is standard Brownian. We assume that the investment strategy $\pi : T \to \mathbb{R}^{n+1}$ is outside the simplex

$$S = \{\pi \in \mathtt{R}^{n+1} : \pi = (\pi^0, \ \ldots, \ \pi^n)^T, \pi^0 + \ldots + \pi^n = 1, \ \pi^0 \geq 0, \ldots \pi^n \geq 0\}.$$

In this case, short selling is possible. The *investment return on bank HQLAs* is then $h : \Omega \times R \to \mathbb{R}^+$, where the dynamics of h can be written as

$$dh_t = \pi_t^T dy_t = \pi_t^T r_t^y dt + \pi_t^T \Sigma_t^y dW_t^y.$$

This notation can be simplified as follows. We denote

$$
\begin{aligned}
r^{\mathrm{R}}(t) &= r^{y^0}(t),\ \ r^{\mathrm{R}} : T \to \mathbb{R}^+, \text{ the rate of return on riskless assets,} \\
r_t^y &= (r^{\mathrm{R}}(t), \widetilde{r_t^y}^T + r^{\mathrm{R}}(t)1_n)^T,\ \widetilde{r}^y : T \to \mathbb{R}^n, \\
\pi_t &= (\pi_t^0, \widetilde{\pi}_t^T)^T = (\pi_t^0, \pi_t^1, \ldots, \pi_t^k)^T,\ \widetilde{\pi} : T \to \mathbb{R}^k, \\
\Sigma_t^y &= \begin{pmatrix} 0 \ldots 0 \\ \widetilde{\Sigma}_t^y \end{pmatrix},\ \widetilde{\Sigma}_t^y \in \mathbb{R}^{n \times n}, \\
\widetilde{C}_t &= \widetilde{\Sigma}_t^y \widetilde{\Sigma_t^y}^T. \text{ Then, we have that} \\
\pi_t^T r_t^y &= \pi_t^0 r^{\mathrm{R}}(t) + \widetilde{\pi}_t^{jT} \widetilde{r}_t^y + \widetilde{\pi}_t^{jT} r^{\mathrm{R}}(t) 1_n = r^{\mathrm{R}}(t) + \widetilde{\pi}_t^T \widetilde{r}_t^y, \\
\pi_t^T \Sigma_t^y dW_t^y &= \widetilde{\pi}_t^T \widetilde{\Sigma}_t^y dW_t^y, \\
dh_t &= [r^{\mathrm{R}}(t) + \widetilde{\pi}_t^T \widetilde{r}_t^y]dt + \widetilde{\pi}_t^T \widetilde{\Sigma}_t^y dW_t^y,\ h(t_0) = h_0.
\end{aligned} \tag{7}
$$

Next, we take $i : \Omega \times T \to \mathbb{R}^+$ as the *increase of NCOs before outflows per monetary unit of NCOs*, $r^i : T \to \mathbb{R}^+$ is the rate of NCO increase before outflows per monetary unit of NCOs, the scalar $\sigma^i \in \mathbb{R}$ is the volatility in the NCO increase before outflows per monetary unit of NCOs and $W^i : \Omega \times T \to \mathbb{R}$ represents standard Brownian motion (compare with [10]). Then, we set

$$
di_t = r_t^i dt + \sigma^i dW_t^i,\ \ i(t_0) = i_0. \tag{8}
$$

The stochastic process i_t in (8) may typically originate from NCOs that have recently taken place or instability in the value of pre-existing NCOs.

Next, we develop a simple stochastic model that replaces a more realistic system that emphasizes features that are specific to our particular study (see, also, [13]). In our situation, we derive models for HQLAs, x^1, and NCOs, x^2, given by

$$
\begin{aligned}
dx_t^1 &= x_t^1 dh_t + x_t^2 u_t^3 dt - x_t^2 de_t \\
&= [r^{\mathrm{R}}(t) x_t^1 + x_t^1 \widetilde{\pi}_t^T \widetilde{r}_t^y + x_t^2 u_t^1 + x_t^2 u_t^2 - x_t^2 r_t^e]dt \\
&\quad + [x_t^1 \widetilde{\pi}_t^T \widetilde{\Sigma}_t^y dW_t^y - x_t^2 \sigma^e dW_t^e],
\end{aligned} \tag{9}
$$

$$
\begin{aligned}
dx_t^2 &= x_t^2 di_t - x_t^2 de_t \\
&= x_t^2 [r_t^i dt + \sigma^i dW_t^i] - x_t^2 [r_t^e dt + \sigma^e dW_t^e] \\
&= x_t^2 [r_t^i - r_t^e]dt + x_t^2 [\sigma^i dW_t^i - \sigma^e dW_t^e].
\end{aligned} \tag{10}
$$

The SDEs (9) and (10) may be rewritten into matrix-vector form in the following way (compare with [10]).

Definition 2.4. (Stochastic System for the LCR Model) *Define the* stochastic system for the LCR model *as*

$$dx_t = A_t x_t dt + N(x_t)u_t dt + a_t dt + S(x_t, u_t)dW_t, \tag{11}$$

with the various terms in this SDE being

$$\begin{aligned}
u_t &= \begin{bmatrix} u_t^2 \\ \widetilde{\pi}_t \end{bmatrix}, \quad u : \Omega \times T \to \mathbb{R}^{n+1}, \\
A_t &= \begin{bmatrix} r^{\mathrm{R}}(t) & -r_t^e \\ 0 & r_t^i - r_t^e \end{bmatrix}; \\
N(x_t) &= \begin{bmatrix} x_t^2 & x_t^1 r_t^{\widetilde{y}^T} \\ 0 & 0 \end{bmatrix}; \; a_t = \begin{bmatrix} x_t^2 u_t^1 \\ 0 \end{bmatrix}; \\
S(x_t, u_t) &= \begin{bmatrix} x_t^1 \widetilde{\pi}_t^T \widetilde{\Sigma}_t^y & -x_t^2 \sigma^e & 0 \\ 0 & -x_t^2 \sigma^e & x_t^2 \sigma^i \end{bmatrix}; \\
W_t &= \begin{bmatrix} W_t^y \\ W_t^e \\ W_t^i \end{bmatrix},
\end{aligned}$$

where W_t^y, W_t^e and W_t^i are mutually (stochastically) independent standard Brownian motions. It is assumed that for all $t \in T$, $\sigma_t^e > 0$, $\sigma_t^i > 0$ and $\widetilde{C}_t > 0$ (compare with [10]).

We can rewrite (11) as follows.

$$\begin{aligned}
N(x_t)u_t &:= \begin{bmatrix} x_t^2 \\ 0 \end{bmatrix} u_t^2 + \begin{bmatrix} x_t^1 r_t^{\widetilde{y}^T} \\ 0 \end{bmatrix} \widetilde{\pi}_t \\
&:= \begin{bmatrix} 0 & 1 \\ 0 & 0 \end{bmatrix} x_t u_t^3 + \sum_{j=1}^{n} \begin{bmatrix} x_t^1 \widetilde{r}_t^{y,j} \\ 0 \end{bmatrix} \widetilde{\pi}_t^j \\
&:= B_0 x_t u_t^0 + \sum_{j=1}^{n} \begin{bmatrix} \widetilde{r}_t^{y,j} & 0 \\ 0 & 0 \end{bmatrix} x_t \widetilde{\pi}_t^j \\
&:= \sum_{j=0}^{n} [B^j x_t] u_t^j;
\end{aligned}$$

and

$$
\begin{aligned}
S(x_t, u_t)dW_t &= \begin{bmatrix} [\widetilde{\pi}_t^T \widetilde{C}_t \widetilde{\pi}_t]^{1/2} & 0 \\ 0 & 0 \end{bmatrix} x_t dW_t^1 + \\
&+ \begin{bmatrix} 0 & -\sigma^e \\ 0 & -\sigma^e \end{bmatrix} x_t dW_t^2 + \begin{bmatrix} 0 & 0 \\ 0 & \sigma^i \end{bmatrix} x_t dW_t^3 \\
&= \sum_{j=1}^{3} [M^j(u_t)x_t] dW_t^j,
\end{aligned}
$$

where $M^j(u_t)$ is the matrix notation used to denote matrices with entries related to u_t. Furthermore, W^1, W^2, and W^3 represent W^y, W^e, and W^i, respectively. From (11) it is evident that $u = (u^2, \widetilde{\pi})$ affects only the SDE of x_t^1 but not that of x_t^2. In particular, for (11) we have that $\widetilde{\pi}$ affects the variance of x_t^1 and the drift of x_t^1 via the term $x_t^1 \widetilde{r_t^y}^T \widetilde{\pi}_t$. On the other hand, u^2 affects only the drift of x_t^1. Then (11) becomes

$$
dx_t = A_t x_t dt + \sum_{j=0}^{n} [B^j x_t] u_t^j dt + a_t dt + \sum_{j=1}^{3} [M^j(u_t)x_t] dW_t^j. \tag{12}
$$

2.2. Description of the simplified LCR model

The model can be simplified if attention is restricted to the system with the LCR, as stated earlier, denoted in this section by $l_t = x_t^1.(x_t^2)^{-1}$ (compare with [10]).

Definition 2.5. (Stochastic Model for a Simplified LCR) *Define the* simplified LCR system *by the SDE*

$$
\begin{aligned}
dx_t &= x_t[r^{\mathrm{R}}(t) + r_t^e - r_t^i + (\sigma^e)^2 + (\sigma^i)^2 + \widetilde{r_t^y}^T \widetilde{\pi}_t]dt \\
&\quad + [u_t^1 + u_t^2 - r_t^e - (\sigma^e)^2]dt \\
&\quad + [(\sigma^e)^2(1 - x_t)^2 + (\sigma^i)^2 x_t^2 + x_t^2 \widetilde{\pi}_t^T \widetilde{C}_t \widetilde{\pi}_t]^{1/2} d\overline{W}_t, \quad x(t_0) = x_0.
\end{aligned} \tag{13}
$$

The model is derived as follows. The starting point is the two-dimensional SDE for $x = (x^1, x^2)^T$ as in the equations (9) and (10). Next, we use the Itô's formula (see, for instance, [15]) to determine

$$
\begin{aligned}
d(x_t^2)^{-1} &= -(x_t^2)^{-2}dx_t^2 + \frac{1}{2}2(x^2)_t^{-3}d < x^2, x^2 > t \\
&= [-(x^2)_t^{-1}(r_t^i - r_t^e) + (x_t^2)^{-1}((\sigma^e)^2 + (\sigma^i)^2)]dt \\
&\quad -(x_t^2)^{-1}\left[0 \; -\sigma^e \; \sigma^i\right] dW_t, \\
dx_t &= x_t^1 d(x_t^2)^{-1} + (x_t^2)^{-1}dx_t^1 + d < x^1, (x^2)^{-1} > t \\
&= [r^{\mathrm{R}}(t)x_t - r_t^e + u_t^1 + u_t^2 + x_t \widetilde{r}_t^y \widetilde{\pi}_t \\
&\quad -x_t[r_t^i - r_t^e] + x_t((\sigma^e)^2 + (\sigma^i)^2) - (\sigma^e)^2]dt \\
&\quad + \left(x_t \widetilde{\pi}_t^T \widetilde{\Sigma}_t^y \; -\sigma^e(1 - x_t) \; -\sigma^i x_t \right) dW_t \\
&= x_t[r^{\mathrm{R}}(t) + r_t^e - r_t^i + (\sigma^e)^2 + (\sigma^i)^2 + \widetilde{r_t^y}^T \widetilde{\pi}_t]dt + \\
&\quad +[u_t^1 + u_t^2 - r_t^e - (\sigma^e)^2]dt \\
&\quad + \left[(\sigma^e)^2(1 - x_t)^2 + (\sigma^i)^2(x_t)^2 + (x_t)^2 \widetilde{\pi}_t^T \widetilde{C}_t \widetilde{\pi}_t\right]^{1/2} d\overline{W}_t,
\end{aligned}
$$

for stochastic $\overline{W} : \Omega \times T \to \mathrm{R}$ that is a standard Brownian motion. Note that in the drift of the SDE (13), the term

$$-r_t^e + x_t r_t^e = -r_t^e(x_t - 1),$$

appears because it models the effect of depreciation of both HQLAs and NCOs. Similarly, the term $-(\sigma^e)^2 + x_t(\sigma^e)^2 = (\sigma^e)^2(x_t - 1)$ appears.

The predictions made by our previously constructed model are consistent with the empirical evidence in contributions such as [13]. For instance, in much the same way as we do, [13] describes how NCOs affect LCRs. On the other hand, to the best of our knowledge, the modeling related to collateral and LCR reference processes (see Section 3 for a comprehensive discussion) have not been tested in the literature before.

3. Optimal Basel III liquidity coverage ratios

In order to determine an optimal cash injection rate (seen as an adjustment to the normal provisioning rate) and HQLA allocation strategy, it is imperative that a well-defined objective function with appropriate constraints is considered. The choice has to be carefully made in order to avoid ambiguous solutions to our stochastic control problem (compare with [10]).

3.1. The optimal bank LCR problem

As in [10], in our contribution, we choose to determine a control law $g(t, x_t)$ that minimizes the cost function $J : \mathcal{G}_A \to \mathrm{R}^+$, where $\mathcal{G}_A$ is the class of admissible control laws

$$\mathcal{G}_A = \{g : T \times \mathcal{X} \to \mathcal{U} | g \text{ Borel measurable} \\ \text{and there exists an unique solution to the closed-loop system}\}, \tag{14}$$

with the closed-loop system for $g \in \mathcal{G}_A$ being given by

$$dx_t = A_t x_t dt + \sum_{j=0}^{n} B^j x_t g^j(t, x_t)dt + a_t dt \\ + \sum_{j=1}^{3} M^j(g(t, x_t))x_t dW_t^j, \quad x(t_0) = x_0. \tag{15}$$

Furthermore, the cost function, $J : \mathcal{G}_A \to \mathbb{R}^+$, of the LCR problem is given by

$$J(g) = \mathbf{E}[\int_{t_0}^{t_1} \exp(-r^f(l - t_0))b(l, x_l, g(l, x_l))dl \\ + \exp(-r^f(t_1 - t_0))b^1(x(t_1))], \tag{16}$$

where $g \in \mathcal{G}_A$, $T = [t_0, t_1]$ and $b^1 : \mathcal{X} \to \mathbb{R}^+$ is a Borel measurable function (compare with [10]). Furthermore, $b : T \times \mathcal{X} \times \mathcal{U} \to \mathbb{R}^+$ is formulated as

$$b(t, x, u) = b^2(u^2) + b^3(x^1/x^2),$$

for $b^2 : \mathcal{U}_2 \to \mathbb{R}^+$ and $b^3 : \mathbb{R}^+ \to \mathbb{R}^+$. Also, $r^f \in \mathbb{R}$ is called the *NCO forecasting rate*, where b^1, b^2 and b^3 are chosen below. In order to clarify the stochastic problem, the following assumption is made.

Assumption 3.1. (Admissible Class of Control Laws) *Assume that* $\mathcal{G}_A \neq \emptyset$.

We are now in a position to state the stochastic optimal control problem for a continuous-time LCR model that we solve (compare with [10]). The said problem may be formulated as follows.

Problem 3.2. (Optimal Bank LCR Problem) *Consider the stochastic system* (15) *for the LCR problem with the admissible class of control laws,* $\mathcal{G}_A$*, given by* (14) *and the cost function,* $J : \mathcal{G}_A \to \mathbb{R}^+$*, given by* (16)*. Solve*

$$\inf_{g \in \mathcal{G}_A} J(g),$$

that amounts to determining the value J^*, given by

$$J^* = \inf_{g \in \mathcal{G}_A} J(g),$$

and the optimal control law g^*, if it exists,

$$g^* = \arg\min_{g \in \mathcal{G}_A} J(g) \in \mathcal{G}_A.$$

3.2. Optimal bank LCRs in the simplified case

In this section, we determine a solution to Problem 3.2 in the case where the term $[t_0, t_1]$ is fixed. In order to find the optimal control processes, we use the dynamic programming algorithm for stochastic optimization where we consider an appropriate Hamilton-Jacobi-Bellman equation (HJBE). In the sequel, we assume that the optimal control laws exist, with the objective function, J, given by (16) being continuous twice-differentiable. Then a combination of integral calculus and Itô's formula (see, for instance, [15]) shows that the value function v satisfies (20) and (21).

Consider the simplified system (13) for the LCR problem with the admissible class of control laws, $\mathcal{G}_A$, given by (14) but with $\mathcal{X} = \mathbb{R}$ (compare with [10]). In this section, we have to solve

$$\begin{aligned} & \inf_{g \in \mathcal{G}_A} J(g), \\ J^* &= \inf_{g \in \mathcal{G}_A} J(g), \qquad (17) \\ J(g) &= \mathbf{E}\Big[\int_{t_0}^{t_1} \exp(-r^f(l - t_0))[b^2(u_t^2) + b^3(x_t)]dt \\ & \quad + \exp(-r^f(t_1 - t_0))b^1(x(t_1))\Big], \qquad (18) \end{aligned}$$

where $b^1 : \mathbb{R} \to \mathbb{R}^+$, $b^2 : \mathbb{R} \to \mathbb{R}^+$ and $b^3 : \mathbb{R}^+ \to \mathbb{R}^+$ are all Borel measurable functions. For the simplified case, the optimal cost function (17) is determined with the simplified cost function, $J(g)$, given by (18). In this case, assumptions have to be made in order to find a solution for the optimal cost function, J^* (compare with [10]). Next, we state an important result about optimal bank coverager ratios in the simplified case.

Theorem 3.3. (Optimal Bank LCRs in the Simplified Case)
Suppose that g^{2*} and g^{3*} are the components of the optimal control law, g^*, that deal with the optimal cash injection rate, u^{2*}, and optimal HQLA allocation, π^{k*}, respectively. Consider the nonlinear optimal stochastic control problem for the simplified LCR system (13) *formulated in Problem* 3.2. *Suppose that the following assumptions hold.*

1. *The cost function is assumed to satisfy*

$$
\begin{aligned}
&b^2(u^2) \in C^2(\mathbb{R}), \\
&\lim_{u^2 \to -\infty} D_{u^2} b^2(u^2) = -\infty, \quad \lim_{u^2 \to +\infty} D_{u^2} b^2(u^2) = +\infty; \\
&D_{u^2 u^2} b^2(u^2) > 0, \ \forall u^2 \in \mathbb{R},
\end{aligned}
\tag{19}
$$

with the differential operator, D, that is applied in this case to function b^2.

2. *There exists a function* $v : T \times \mathbb{R} \to \mathbb{R}, \ \ v \in C^{1,2}(T \times \mathcal{X})$, *that is a solution of the HJBE given by*

$$
\begin{aligned}
0 = {} & D_t v(t,x) + \frac{1}{2}[(\sigma^e)^2(1-x)^2 + (\sigma^i)^2(x_t)^2] D_{xx} v(t,x) \\
& + x(r^{\mathbb{R}}(t) + r_t^e - r_t^i + (\sigma^e)^2 + (\sigma^i)^2) D_x v(t,x) \\
& + [u_t^1 - r_t^e - (\sigma^e)^2] D_x v(t,x) \\
& + u_t^{2^*} D_x v(t,x) + \exp(-r^f(t-t_0)) b^2(u_t^{2^*}) \\
& + \exp(-r^f(t-t_0)) b^3(x) - \frac{[D_x v(t,x)]^2}{2 D_{xx} v(t,x)} \widetilde{r}_t^{y\,T} \widetilde{C}_t^{-1} \widetilde{r}_t^y,
\end{aligned}
\tag{20}
$$

$$
v(t_1, x) = \exp(-r^f(t_1 - t_0)) b^1(x), \tag{21}
$$

where u^{2*} *is the unique solution of the equation*

$$
0 = D_x v(t,x) + \exp(-r^f(t-t_0)) D_{u^2} b^2(u_t^2). \tag{22}
$$

Then the optimal control law is

$$
\begin{aligned}
g^{2*}(t,x) &= u^{2^*}, \ g^{2^*} : T \times \mathcal{X} \to \mathbb{R}^+, \qquad \text{(23)} \\
&\quad \textit{with } u^{2*} \in \mathcal{U}_2 \textit{ the unique solution of the equation (22)} \\
\widetilde{\pi}^* &= -\frac{D_x v(t,x)}{x D_{xx} v(t,x)} \widetilde{C}_t^{-1} \widetilde{r}_t^y, \qquad \text{(24)} \\
g^{3,k*}(t,x) &= \min\{1, \max\{0, \widetilde{\pi}^{k*}\}\}, g^{3,k*} : T \times \mathcal{X} \to \mathbb{R}, \qquad \text{(25)}
\end{aligned}
$$

Furthermore, the value of the problem is

$$
J^* = J(g^*) = \mathbf{E}[v(t, x_0)]. \tag{26}
$$

Next, we choose a particular cost functions for which an analytic solution can be obtained for the value function and control laws (compare with [10]). The following theorem provides the optimal control laws for quadratic cost functions.

Theorem 3.4. (Optimal Bank LCRs with Quadratic Cost Functions): *Consider the nonlinear optimal stochastic control problem for the simplified LCR system* (13) *formulated in Problem* 3.2. *Consider the cost function*

$$\begin{aligned} J(g) = \mathbf{E}\Big[\int_{t_0}^{t_1} \exp(-r^f(l-t_0))[\tfrac{1}{2}c^2((u^2)^2(l)) + \tfrac{1}{2}c^3(x_t - l^r)^2]dl \\ + \frac{1}{2}c^1(x(t_1) - l^r)^2 \exp(-r^f(t_1 - t_0))\Big]. \end{aligned} \quad (27)$$

We assume that the cost functions satisfy

$$\begin{aligned} b^1(x) &= \frac{1}{2}c^1(x - l^r)^2, \;\; c^1 \in (0, \infty); \\ b^2(u^2) &= \frac{1}{2}c^2(u^2)^2, \;\; c^2 \in (0, \infty); && (28) \\ b^3(x) &= \frac{1}{2}c^3(x - l^r)^2, \;\; c^3 \in (0, \infty), && (29) \\ & l^r \in \mathbb{R}, \text{ called the reference value of the LCR} \end{aligned}$$

Define the first-order ODE

$$\begin{aligned} -\dot{q}_t &= -(q_t)^2/c^2 + c^3 + q_t 2(r^{\mathrm{R}}(t) + r^e_t - r^i_t + (\sigma^e)^2 + (\sigma^i)^2) \\ &\quad + q_t[-r^f - {r^{\widetilde{y}}_t}^T \widetilde{C}_t^{-1} \widetilde{r}^y_t + (\sigma^e)^2 + (\sigma^i)^2], \;\; q_{t_1} = c^1; && (30) \\ -\dot{x^r}_t &= -c^3(x^r_t - l^r)/q_t - x^r_t[r^{\mathrm{R}}(t) + r^e_t - r^i_t + (\sigma^e)^2 + (\sigma^i)^2] \\ &\quad - [u^1_t - r^e_t - (\sigma^e)^2] - (x^r_t - 1)((\sigma^e)^2 + (\sigma^i)^2) - (\sigma^i)^2, \;\; x^r(t_1) = l^r; && (31) \\ -\dot{l}_t &= -r^f l_t + c^3(x^r_t - l^r)^2 - q_t(\sigma^e)^2(x^r_t - 1)^2 - q_t(\sigma^i)^2(x^r_t)^2, \\ &\quad l(t_1) = 0. && (32) \end{aligned}$$

The function $x^r : T \to \mathbb{R}$ *will be called the* LCR reference (process) function. *Then we have that the following hold.*

(a) *There exist solutions to the ordinary differential equations* (30), (31) *and* (32). *Moreover, for all* $t \in T$, $q_t > 0$.

http://dx.doi.org/10.5772/58395

(b) The optimal control laws are

$$u^{2*}_t = -(x - x^r_t)q_t/c^2, \tag{33}$$
$$g^{2*}(t,x) = u^{2*},\ g^{2*} : T \times \mathcal{X} \to \mathbb{R}^+,$$
$$\widetilde{\pi}^*_t = -\frac{(x - x^r_t)}{x}\widetilde{C}^{-1}_t \widetilde{r}^y_t,\ g^{3*} : T \times \mathcal{X} \to \mathbb{R}^k, \tag{34}$$
$$g^{3,k*}(t,x) = \begin{cases} \widetilde{\pi}^{k*}, & \text{if } \widetilde{\pi}^{k*} \in [0,1], \\ \min\{1, \max\{0, \widetilde{\pi}^{k*}_t\}\}, & \text{otherwise}, \end{cases} \quad \forall k \in \mathbb{Z}^n. \tag{35}$$

(c) The value function and the value of the problem are

$$v(t,x) = \exp(-r^f(t_1 - t_0))[\frac{1}{2}(x - x^r_t)^2 q_t + \frac{1}{2}l_t], \tag{36}$$
$$J^* = J(g^*) = \mathbf{E}[v(t_0, x_0)]. \tag{37}$$

For the cost on the cash injection, the function (28) is considered, where the input variable u^2 is restricted to the set $\mathbb{R}^+$. If $u^2 > 0$ then the banks should acquire additional HQLAs. The cost function should be such that cash injections are maximized, hence $u^2 > 0$ should imply that $b^2(u^2) > 0$. For Theorem 3.4 we have selected the cost function $b^2(u^2) = \frac{1}{2}c^2(u^2)^2$, given by (28). Here, both positive and negative values of u^2 are penalized equally. An important reason for this is that an analytic solution of the value function can be determined (compare with [10]).

The reference process, l^r, may be 1 that is the threshold for the LCR standard. The cost on meeting liquidity provisioning will be encoded in a cost on the LCR. If the LCR, $x > l^r$, is strictly larger than a set value l^r, then there should be a strictly negative cost. If, on the other hand, $x < l^r$, then there may be a positive cost. We have selected the cost function $b^3(x) = \frac{1}{2}c^3(x - l^r)^2$ in Theorem 3.4 given by (29). This is also done to obtain an analytic solution of the value function and that case by itself is interesting (see, for instance, [10] for more details). Another cost function that we can consider is

$$b^3(x) = c^3[\exp(x - l^r) + (l^r - x) - 1],$$

that is strictly convex and asymmetric in x with respect to the value l^r. For this cost function, it is reasonable that costs with $x > l^r$ are penalized lower than those with $x < l^r$. Another cost function considered is to keep $b^3(x) = 0$ for $x < l^r$ (see, for instance, [10]).

4. Numerical results for LCRs

In this section, we provide numerical-quantitative results about LCRs and their connections with HQLAs and NCOs to supplement the theoretical-quantitative treatment in Sections 2 and 3 (see [13] for more details). More precisely, we describe the LCR data and descriptive statistics for Class I and II banks for the sample period 2002 to 2012.

4.1. Description of banking data

In this subsection, we describe the banking data pertaining to LCRs.

4.1.1. Class I and II banks

We investigate liquidity for Class I banks that hold more than US $ 4 billion in Tier 1 capital (T1K) and are internationally active. Moreover, we consider Class II banks that violate one or both of these conditions (see, for instance, [9] and [17]). In reality, some Class II banks considered could have been classified as Class I if they were internationally active. Nevertheless, these banks make a large contribution to the total assets of Class II banks. Invariably, all Class I banks can also be classified as large in that their gross total assets (GTA) exceed US $ 3 billion. Many of the banks in our study come from jurisdictions affiliated to the BCBS and Macro-Economic Assessment Group (MAG).

Our investigation includes 157 Class I and 234 Class II LIBOR-based banks from 38 countries. These banks (with the number of Class I and Class II banks in parenthesis for each jurisdiction, as well as * and ′ denoting BCBS and MAG members, respectively) are located in Argentina* (1,3), Australia*′ (5,2), Austria (2,5), Belgium* (1,2), Botswana (1,1), Brazil*′ (3,1), Canada*′ (7,3), China*′ (7,1), Czech Republic (4,3), Finland (0,14), France*′ (5,5), Germany*′ (7,24), Hong Kong SAR* (1,8), Hungary (1,2), India* (6,6), Indonesia* (1,3), Ireland (3,1), Italy*′ (2,11), Japan*′ (14,5), Korea*′ (6,4), Luxembourg* (0,1), Malta (0,3), Mexico*′ (1,8), Namibia (0,1), the Netherlands*′ (3,13), Norway (1,6), Poland (0,5), Portugal (3,3), Russia* (0,3), Saudi Arabia* (4,1), Singapore* (5,0), South Africa* (4,5), Spain*′ (2,4), Sweden* (4,0), Switzerland*′ (3,5), Turkey* (7,1), United Kingdom*′ (8,5) and United States*′ (35,66). In order to limit depositor losses, all 38 jurisdictions have explicit deposit insurance schemes or implicit government protection schemes for banks.

4.1.2. Banking data restrictions

In our study, we did not consider Central Banks, subsidiaries, banks with incomplete (inconsistent or non-continuous) information nor observations with negative HQLA, NCO, ASF, RSF or other values (see, for instance, [9] and [17]). Furthermore, we use non-permanent samples that do not suffer from survivorship bias to study cross sectional patterns. For our sample, bank failure data for the period 2002 to 2012 was obtained from deposit insurance schemes or implicit government protection schemes. For instance, for the US, such data was obtained from the Federal Deposit Insurance Corporation (see [9] and [17] for more details). We choose the period 2002-2012 because available EMERG global liquidity data does not allow us to reliably determine the LCR and NSFR prior to 2002 (see, for instance, [9] and [17]).

4.1.3. Banking data computations

Estimating the LCR and NSFR using available EMERG public data proved to be a challenge. Firstly, the prescripts for these risk standards are sometimes ambiguous and subject to frequent regulatory amendment. For instance, the final rules relating to the LCR were only published on Monday, 7 January 2013.

Secondly, the EMERG global banking data has several limitations in terms of granularity and format when compared with the information required to determine the Basel III liquidity standards (see, for instance, [9] and [17]). In all instances, we had to make difficult choices when applying Basel III guidelines to such a large diversity of banks.

In the absence of suitable data, we were heavily dependent on the interpolation and extrapolation techniques discussed below. Firstly, it is clear that the LCR calculation requires information about liabilities with a remaining maturity of less than 1 month. However, quarterly EMERG data provides information about liabilities with a remaining maturity of less than 3 months. So we had to extrapolate the liabilities with a remaining maturity of 1 month. There are two approaches to doing this. In the first instance, we can assume the maturity schedule is evenly distributed, such that the amount of liabilities with a remaining maturity of less than 1 month equals 1/3 of the amount of liabilities with a remaining maturity of less than 3 months. This is the approach adopted in this chapter. Secondly, as a robustness check, we can assume an extreme case, such that all liabilities with a remaining maturity within 3 months mature within the first month. In this instance, the guidelines require dividing liabilities into subcategories of retail deposits, unsecured wholesale funding and secured funding with different run-off rates (see, for instance, [9] and [17]). However, the information available from the EMERG global data lacks such granularity. Out of necessity, we have to make assumptions on the distribution of subcategories within their primary category. Without additional information, we generally assume equal distribution of subcategories within the primary category , . Finally, except for unused commitments, letters of credit and the net fair value of derivatives, we do not have the information required for calculating the liquidity needs of all other OBS items, such as increased liquidity needs related to downgrade triggers embedded in financing transactions, derivatives and other contracts. Therefore, our calculations of the LCR and NSFR are partial measures that capture a bank's liquidity risk as mainly reflected by its BS and to a lesser extent its OBS items (see [9] and [17] for more information).

4.2. 2002 to 2012 LCRs for class I and II banks

In this subsection, we provide 2002 to 2012 LCRs for Class I and II banks.

Table 1 shows that the LCR has been in a downward trend from 2002 through 2007. The average LCR had risen sharply from 2007 to 2009 and peaked in 2009. The general impression from Figure 1 is that the LCR time series is non-stationary.

4.3. Descriptive statistics for LCRs of class I and II banks

In this subsection, we provide 2002 to 2012 LCR descriptive statistics for Class I and II banks.

Table 2 reports the summary statistics of the approximate measures of the LCR for Class I banks, where the mean for the LCR is 74.96 %. In this table, the LCR displays positive skewness. The value of the kurtosis for the LCR in Table 2 is equal to or less than 3, that means that the distribution is flat. The LCR risk measure exhibits normality because the p-values are greater than 5 %. Nevertheless, the normality test is very sensitive to the number of observations and may only produce desirable and efficient results if observations are large. From Table 2, it is clear that, in the absence of empirical evidence, it is hard to conclude that the Basel III LCR standard had complied with these standards.

Bank Class			Bank Class		
Quarter	**Class I**	**Class II**	**Quarter**	**Class I**	**Class II**
02Q1	0.81832	0.84588	02Q2	0.79516	0.82194
02Q3	0.8264	0.85386	02Q4	0.86464	0.89376
03Q1	0.78884	0.77406	03Q2	0.74112	0.76608
03Q3	0.7334	0.7581	03Q4	0.71024	0.73416
04Q1	0.6562	0.6783	04Q2	0.62532	0.64638
04Q3	0.63304	0.65436	04Q4	0.6176	0.6384
05Q1	0.62532	0.64638	05Q2	0.6176	0.6384
05Q3	0.60988	0.63042	05Q4	0.6176	0.6384
06Q1	0.59444	0.61446	06Q2	0.57128	0.59052
06Q3	0.57128	0.59052	06Q4	0.54812	0.56658
07Q1	0.5404	0.5586	07Q2	0.579	0.5985
07Q3	0.55584	0.51456	07Q4	0.54812	0.56658
08Q1	0.72568	0.75012	08Q2	0.74884	0.77406
08Q3	0.74884	0.77406	08Q4	0.74412	0.76608
09Q1	0.88008	0.90972	09Q2	0.95728	1.08114
09Q3	1.02676	1.06134	09Q4	0.91566	0.99753
10Q1	0.93412	0.96558	10Q2	0.95376	0.96184
10Q3	0.96552	0.92568	10Q4	0.97868	0.94962
11Q1	0.98324	0.93366	11Q2	0.99288	0.91566
11Q3	0.99692	0.98578	11Q4	1.02391	1.08113
12Q1	1.07414	1.16982	12Q2	1.10664	1.19376
12Q3	1.16849	1.23778	12Q4	1.19281	1.29002

Table 1. 2002 to 2012 LCRs for Class I and II Banks

Descriptive Statistics of LCR for Class I and II Banks		
Parameter	**Class I LCR**	**Class II LCR**
Mean	0.748720	0.773430
Median	0.748840	0.774060
Maximum	1.026760	1.061340
Minimum	0.540400	0.514560
Std. Dev.	0.027670	0.143466
Skewness	0.027670	-0.02945
Kurtosis	1.825270	1.855696
Jarque-Bera	2.535599	2.406985
Probability	0.281450	0.300144
Sum	32.94368	34.03091
Sum Sq. Dev.	0.794993	0.885043
Observations	44	44

Table 2. Descriptive Statistics of LCR for Class I and II Banks

5. Conclusions and future directions

In this section, we draw conclusions about the LCR modeling and optimization and related numerical examples. Furthermore, we suggest possible topics for future research.

5.1. Conclusions

In this subsection, we make conclusions about the LCR model, optimal Basel III LCRs and numerical results for LCRs.

5.1.1. Conclusions about the LCR model

One of the main contributions of this book chapter is the way the LCR dynamics model is constructed by using stochastic techniques. This model depends on HQLAs, NCOs as well as the liquidity provisioning rate. We believe that this is an addition to pre-existing literature because it captures some of the uncertainty associated with LCR variables. In this regard, we provide a theoretical-quantitative modeling framework for establishing bank LCR reference processes and the making of decisions about liquidity provisioning rates and asset allocation.

In Subsecdtion 2.1, we mention the possibility of adjusting the cash injection rate depending on whether the bank is experiencing deficit or surplus liquidity. The latter occurs where cashflows into the banking system persistently exceed withdrawals of liquidity from the market by the central bank. This is reflected in holdings of reserves in excess of the central bank's required reserves. Transitional economies, for example, often attract large capital inflows as the economy opens and undergoes privatization. The effect of these inflows on liquidity is often magnified by central bank intervention in the foreign exchange market when there is upward pressure on the domestic currency. In the wartime economy, consumption is restricted and large amounts of involuntary savings accumulate until goods and services eventually become more widely available. Soviet-style economies have displayed widespread shortages and administered prices. This creates a situation of repressed inflation, whereby prices are too low relative to the money stock, leaving individuals with excess real balances. The importance of surplus liquidity for central banks is threefold and lies in its potential to influence: (1) the transmission mechanism of monetary policy; (2) the conduct of central bank intervention in the money market, and (3) the central bank's balance sheet and income.

5.1.2. Conclusions about optimal basel III LCRs

We obtained an analytic solution to an optimal bank LCR problem with a quadratic objective function. In principle, this solution can assist in managing LCRs. Here, liquidity provisioning and HQLA allocation are expressed in terms of a reference process. To our knowledge such processes have not been considered for LCRs before. This chapter makes a clear connection between liquidity and financial crises in a numerical-quantitative framework.

An interpretation of the control laws given by (33) and (34) follows. In times of deficit, the cash injection rate, u^{2^*}, is proportional to the difference between the LCR, x, and the reference process for this ratio, x^r. The proportionality factor is q_t/c^2 that depends on the relative ratio of the cost function on u^2 and the deviation from the reference ratio, $(x - x^r)$. The property that the control law is symmetric in x with respect to the reference process x^r is a direct consequence of the cost function $b^r(x) = \frac{1}{2}c^3(x - x^r)^2$ being symmetric with respect to $(x - x^r)$. The optimal portfolio distribution is proportional to the relative difference between the LCR and its reference process, $(x - x_t^r)/x$. This seems natural. The proportionality factor is $\widetilde{C}_t^{-1}\widetilde{r}_t^y$ that represents the relative rates of asset return multiplied with the inverse of the corresponding variances. It is surprising that the control law has this structure. Apparently the optimal control law is not to liquidate first all HQLAs with the highest liquidity provisioning rate, then the HQLAs with the next to highest liquidity provisioning rate, etc. The proportion of all HQLAs depend on the relative weighting in $\widetilde{C}_t^{-1}\widetilde{r}_t^y$ and not on the deviation $(x - x_t^r)$.

The novel structure of the optimal control law is the *LCR reference process*, $x^r : T \to \mathbb{R}$. The differential equation for this reference function is given by (31). This equation is new for the area of LCR control and therefore deserves discussion. The differential equation has several terms on its right-hand side that will be discussed separately. Consider the term

$$u_t^1 - r_t^e - (\sigma^e)^2.$$

This represents the difference between normal rate of liquidity provisioning per monetary unit of the bank's NCOs and NCO outflows, where r_t^e is the rate of outflow per monetary unit of NCOs. Note that if $[u_t^1 - r_t^e - (\sigma^e)^2] > 0$, then the reference LCR function can be increasing in time due to this inequality so that, for $t > t_1$, $x_t < l^r$. The term $c^3(x_t^r - l^r)/q_t$ models that if the reference LCR function is smaller than l^r, then the function has to increase with time. The quotient c^3/q_t is a weighting term that accounts for the running costs and for the effect of the solution of the Riccati differential equation. The term

$$x_t^r[r^{\mathrm{R}}(t) + r_t^e - r_t^i + (\sigma^e)^2 + (\sigma^i)^2],$$

accounts for two effects. The difference $r_t^e - r_t^i$ is the nett effect of the rate of outflows per monetary unit of the bank's NCOs, r^e, and rate of NCO increase before outflows per monetary unit of NCOs, r_t^i. The term $r^{\mathrm{R}}(t) + (\sigma^e)^2 + (\sigma^i)^2$ is the effect of NCO increase due to the reserves and the variance of riskier liquidity provisioning. The last term

$$(x_t^r - 1)((\sigma^e)^2 + (\sigma^i)^2) - (\sigma^i)^2,$$

accounts for the effect on HQLAs and NCOs. More information is obtained by streamlining the ODE for x^r. In order to accompolish this it is necessary to assume the following.

Assumption 5.1. (Liquidity Parameters): *Assume that the parameters of the problem are all time-invariant and also that q has become constant with value q^0.*

Then the differential equation for x^r can be rewritten as

$$
\begin{aligned}
-\dot{x^r}_t &= -k(x_t^r - m),\ x^r(t_1) = l^r; \\
k &= (r^{\mathrm{R}} + r^e - r^i + 2((\sigma^e)^2 + (\sigma^i)^2)) + c^3/q^0; \\
m &= \frac{l^r c^3/q^0 - (u^1 - r^e - (\sigma^e)^2) + (\sigma^e)^2}{(r^{\mathrm{R}} + r^e - r^i + 2((\sigma^e)^2 + (\sigma^i)^2)) + c^3/q^0}.
\end{aligned}
$$

Because the finite horizon is an artificial phenomenon to make the optimal stochastic control problem tractable, it is of interest to consider the long term behavior of the LCR reference trajectory, x^r. If the values of the parameters are such that $k > 0$ then the differential equation with the terminal condition is stable. If this condition holds then $\lim_{t\downarrow 0} q_t = q^0$ and $\lim_{t\downarrow 0} x_t^r = m$ where the down arrow prescribes to start at t_1 and to let t decrease to 0. Depending on the value of m, the control law for at a time very far away from the terminal time becomes then,

$$
\begin{aligned}
u_t^{2*} &= -(x_t - m)q^0/c^2 = \begin{cases} > 0, \text{ if } x_t < m, \\ < 0, \text{ if } x_t > m, \end{cases} \\
\pi_t^* &= -\frac{(x_t - m)}{x_t}\widetilde{Cr}^j = \begin{cases} > 0, \text{ if } x_t < m, \\ < 0, \text{ if } x_t > m, \text{ if } \pi^* < 0 \text{ then set } \pi^* = 0, \end{cases}
\end{aligned}
$$

The interpretation for the two cases follows below.

Case 1 ($x_t > m$): Then the LCR x is too high. This is penalized by the cost function hence the control law prescribes not to invest in riskier HQLAs. The payback advice is due to the quadratic cost function that was selected to make the solution analytically tractable. An increase in liquidity provisioning will increase NCOs that, in turn, will lower the LCR.

Case 2 ($x_t < m$): The LCR x is too low. The cost function penalizes and the control law prescribes to invest more in riskier HQLAs. In this case, more funds will be available and credit risk on the balance sheet will decrease. Thus higher valued HQLAs should be held. On the other hand, when banks hold less HQLAs, they should decrease their NCOs that may lead to higher LCRs.

5.1.3. Conclusions about numerical results for LCRs

We approximate the Basel III standard, LCR, that is a measure of asset liquidity for global EMERG banking data mentioned earlier This is a challenging task given the nature of the data available and the ever-changing nature of Basel III liquidity regulation. In the light of the determined results, our analysis gives us a new understanding of the problem of approximating liquidity risk measures. From Table 1, we observe that from Q209 to Q412 there was a steady increase in the LCR. This is probably due to banks holding more liquid assets and restricting cash outflows and risky activities.

In this paragraph, we highlight how our research on approximating Basel III and traditional liquidity risk measures has advanced the knowledge in this field of endeavor. For both Class I and II banks, our research approximates LCRs for a large diversity of banks for an extended *that amounts to determining the value J^*, given by*

$$J^* = \inf_{g \in \mathcal{G}_A} J(g),$$

and the optimal control law g^*, if it exists,

$$g^* = \arg\min_{g \in \mathcal{G}_A} J(g) \in \mathcal{G}_A.$$

3.2. Optimal bank LCRs in the simplified case

In this section, we determine a solution to Problem 3.2 in the case where the term $[t_0, t_1]$ is fixed. In order to find the optimal control processes, we use the dynamic programming algorithm for stochastic optimization where we consider an appropriate Hamilton-Jacobi-Bellman equation (HJBE). In the sequel, we assume that the optimal control laws exist, with the objective function, J, given by (16) being continuous twice-differentiable. Then a combination of integral calculus and Itô's formula (see, for instance, [15]) shows that the value function v satisfies (20) and (21).

Consider the simplified system (13) for the LCR problem with the admissible class of control laws, $\mathcal{G}_A$, given by (14) but with $\mathcal{X} = \mathbb{R}$ (compare with [10]). In this section, we have to solve

$$\begin{aligned} &\inf_{g \in \mathcal{G}_A} J(g), \\ J^* &= \inf_{g \in \mathcal{G}_A} J(g), \qquad (17) \\ J(g) &= \mathbf{E}\Big[\int_{t_0}^{t_1} \exp(-r^f(l - t_0))[b^2(u_t^2) + b^3(x_t)]dt \\ &\quad + \exp(-r^f(t_1 - t_0))b^1(x(t_1))\Big], \qquad (18) \end{aligned}$$

where $b^1 : \mathbb{R} \to \mathbb{R}^+$, $b^2 : \mathbb{R} \to \mathbb{R}^+$ and $b^3 : \mathbb{R}^+ \to \mathbb{R}^+$ are all Borel measurable functions. For the simplified case, the optimal cost function (17) is determined with the simplified cost function, $J(g)$, given by (18). In this case, assumptions have to be made in order to find a solution for the optimal cost function, J^* (compare with [10]). Next, we state an important result about optimal bank coverager ratios in the simplified case.

Theorem 3.3. (Optimal Bank LCRs in the Simplified Case)
Suppose that g^{2*} and g^{3*} are the components of the optimal control law, g^*, that deal with the optimal cash injection rate, u^{2*}, and optimal HQLA allocation, π^{k*}, respectively. Consider the nonlinear optimal stochastic control problem for the simplified LCR system (13) formulated in Problem 3.2. Suppose that the following assumptions hold.

availability of more suitable data of sufficient granularity as well as improved extrapolation and interpolation techniques.

We have already made several contributions in support of the endeavors outlined in the previous paragraph. For instance, our journal article [12] deals with issues related to liquidity risk and the financial crisis. Also, the role of information asymmetry in a subprime context is related to the main hypothesis of the book [14].

Author details

J. Mukuddem-Petersen[1], M.A. Petersen[1] and MP. Mulaudzi*[2]

*Address all corespodence to mulaump@unisa.ac.za

Faculty of Commerce and Administration, North West University, South Africa

Department of Decision Sciences, University of South Africa

References

[1] Basel Committee on Banking Supervision (January 2013). Basel III: The liquidity coverage ratio and liquidity risk monitoring tools. Bank for International Settlements (BIS) Publications. Retrieved Monday, 25 March 2013 (http://www.bis.org/publ/bcbs238.htm).

[2] Basel Committee on Banking Supervision (October 2011). Progress report on Basel III implementation. Bank for International Settlements (BIS), Basel, Switzerland. Retrieved April 25, 2012, from http://www.bis.org/publ/bcbs193.htm.

[3] Basel Committee on Banking Supervision (BCBS). (June 2011). Basel III: A global regulatory framework for more resilient banks and banking systems – A revised version June 2011. June 2011. Bank for International Settlements (BIS) Publications. Retrieved Monday, 25 March 2013 (http://www.bis.org/publ/bcbs189.htm).

[4] Basel Committee on Banking Supervision (December 2010). Basel III: International framework for liquidity risk measurement, standards and monitoring. Bank for International Settlements (BIS), Basel, Switzerland. Retrieved April 25, 2012, from http://www.bis.org/publ/bcbs188.htm.

[5] Basel Committee on Banking Supervision (September 2008). Principles for sound liquidity risk management and supervision. Bank for International Settlements (BIS), Basel, Switzerland. Retrieved April 25, 2012, from http://www.bis.org/publ/bcbs144.htm.

[6] Demyanyk Y, van Hemert O. Understanding the subprime mortgage crisis. *Review of Financial Studies*, 2011; 24(6), 1848-1880.

[7] Elyasiani E, Mester LJ, Pagano MS, Large Capital Infusions, Investor Reactions, and the Return and Risk-Performance of Financial Institutions Over the Business Cycle (May 3,

2013). FRB of Philadelphia Working Paper No. 13-23. Retrieved January 28, 2014, from http://ssrn.com/abstract=2269223 or http://dx.doi.org/10.2139/ssrn.2269223

[8] Gorton GB. Information, liquidity, and the (ongoing) panic of 2007, American Economic Review, Papers and Proceedings, vol. 99, no. 2 (May 2009), 567-572.

[9] Hong H, Huang J, Wu D. The information content of Basel III liquidity risk measures. Retrieved Monday, 2 December 2013, from: http://ssrn.com/abstract=2177614.

[10] Mukuddem-Petersen J, Petersen MA (2008). Optimizing Asset and Capital Adequacy Management in Banking, Journal of Optimization Theory and Applications 137(1), 205-230.

[11] Petersen MA, De Waal B, Hlatshwayo LNP, Mukuddem-Petersen J (2013). A Note on Basel III and Liquidity. Applied Economic Letters 20(8), 777–780.

[12] Petersen MA, De Waal B, Mukuddem-Petersen J, Mulaudzi MP. Subprime mortgage funding and liquidity risk, *Quantitative Finance*, doi: 10.1080/14697688.2011.637076, 2012.

[13] Petersen MA, Mukuddem-Petersen J. Basel III Liquidity Regulation and Its Implications, Business Expert Press, McGraw-Hill, New York, 2014, ISBN: 978-1-60649-872-9 (print); ISBN: 978-1-60649-873-6 (ebook).

[14] Petersen MA, Senosi MC, Mukuddem-Petersen J. Subprime Mortgage Models, New York: Nova, ISBN: 978-1-61728-694-0, 2012.

[15] Soner HM, Fleming WH. (2008). Controlled Markov Processes and Viscosity Solutions, 2nd Edition, New York: Springer Verlag.

[16] van den End JW, Tabbae M. When liquidity risk becomes a systemic issue: Empirical evidence of bank behaviour, *Journal of Financial Stability*, doi:10.1016/j.jfs.2011.05.003, 2012.

[17] Wu D, Hong H. The information value of Basel III liquidity risk measures. Retrieved Friday, 29 November 2013, from: http://ssrn.com/abstract=2177614.

Chapter 4

Risk-Constrained Forward Trading Optimization by Stochastic Approximate Dynamic Programming

Miguel Gil-Pugliese and Fernando Olsina

Additional information is available at the end of the chapter

http://dx.doi.org/10.5772/57466

1. Introduction

Since the mid-twentieth century, Dynamic Programming (DP) has proved to be a flexible and powerful approach to address optimal decisions problems. Nevertheless, a decisive drawback of the conventional DP is the need for exploring the whole state space in order to find the optimal solution. The immense amount of mathematical operations involved to solve real-scale problems, constrained the application of DP to small or highly simplified cases. Indeed, state space grows exponentially with the number of variables when considering multivariate optimization. The curse of dimensionality is a well-known limitation of conventional DP algorithms for tackling large-scale problems ubiquitous in real science and engineering applications.

In the last decades, many new algorithms emerged in different branches of science to overcome the inherent limitations of conventional DP. Unlike conventional DP, these algorithms avoid enumerating and calculating every possible state of a system during the optimization process. Instead, they estimate relevant features of the state space. This approach circumvents the dimensionality limitations of the conventional DP while retaining many of its advantages.

In this chapter, the application of advanced stochastic dynamic programming techniques to the optimization of the forward sell strategy of a power generator subjected to delivery risk is considered. The proposed approach allows rebalancing the portfolio during the period of analysis. In electricity markets, a power generator can sell in advance part or all its future energy production at a fixed price, hedging against the high price volatility of the spot market. The strategy of eliminating the price risk by selling in advance the entire production in the forward market to a fixed price is often thought as the minimum-risk trading policy. Nonetheless, it can be proven that this is not the case for most generators. The outages of the generation units and transmission lines, as well as unforeseen limitations in the primary energy supply expose generators to delivery risk [1]. Delivery risk considerably modifies the probability distribution of profits, shifting the optimal trading strategy toward a portfolio mixing forward contracts and power sold in the spot market. Because of the size of the probability state space and the limited computing capabilities, the problem of the optimal trading strategy has not a closed form solution and thus, its determination is matter of current study. The increase in computing power and recent developments in Operational Research has brought new insights into the solution of such problems.

In the past decade and by virtue of the ever increasing computational power, many methods emerged in different scientific fields with several different names: Reinforced Learning, Q-Learning, Neuro-Dynamic Programming, etc. All these methods were later brought together in what is currently known as Approximated Dynamic Programming (ADP) [2],[3]. These algorithms resign the exhaustive enumeration and calculation of the space-state typically performed by conventional DP. Instead, they iteratively approximate a function of the space state through stochastic simulation and statistical regression techniques, circumventing the dimensionality problem of DP.

Although ADP algorithms are being used in several other fields of science, the application to design optimal trading strategies in power markets has not been proposed so far. In this chapter, ADP techniques are exploited to optimize the selling strategy of a power generator trading in a frictional market with transaction costs. Three available products are considered: selling in the spot market, and/or get involved in quarterly and one-year forward contracts. The objective of the generator is to maximize the expected profit while limiting financial risk. Decisions can be made only at the beginning of each month. At each decision stage, the current trading position can be changed at a cost in order to rebalance the portfolio.

2. Approximate dynamic programming

In the field of decision making, it is often useful to assess what could be expected from each possible decision given the information available. After evaluating the outcome of each alternative decision, a simple comparison is enough to take the optimal course of action. This approach is straightforward but also naive. Real problems often present simply too many possible options to evaluate. Moreover, if the problem involves sequential decision stages, the number of possible solution paths scales up exponentially. Finally, outcomes are frequently subjected to uncertainty. So, several outcomes could present themselves for each decision, augmenting further the size of the problem.

Therefore, in order to keep the size of the problem within reach, shortcuts and simplifications areoften necessary. Dynamic Programming (DP) is a clever way to reduce the number of options based on Bellman´s Principle of Optimality: "An optimal policy has the property that whatever the initial state and initial decision are, the remaining decisions must constitute an optimal policy with regard to the state resulting from the first decision" [4]. This leads to the fact that from any given state there is an optimal decision trajectory that once solved can be used for every other path containing that state. Therefore for each state is only necessary to hold the information of the best solution until the end, ruling out suboptimal options. This rule prevents the exponential growth in the decision sequence path, scaling the problem only linearly.

Yet, other reasons of explosive dimensionality growth remain, namely the number of states, as well as decision and outcome spaces. For small financial decision problems, conventional DP algorithms are able to find the optimal policy. For real-scale problems, gross simplifications are often necessary to keep tractability. Sometimes, these simplifications render the model unrealistic turning results meaningless.

Finding an appropriate combination of financial instruments in a portfolio can be portrayed as a Markov Decision Problem (MDP). A MDP is defined as a succession of states that are reached through decisions (actions) followed by an outcome (reaction) of the system. After each decision, the system evolves into another state, according probabilities defined by the previous state and the decision taken. Each transition between states has a cost, usually dependent on the action taken, and a reward produced by the reaction of the system, as depicted in Figure 1. In these processes, a decision maker should choose a sequence of actions so that the expected sum of rewards minus the sum of incurred costs is maximized over a certain period of time.

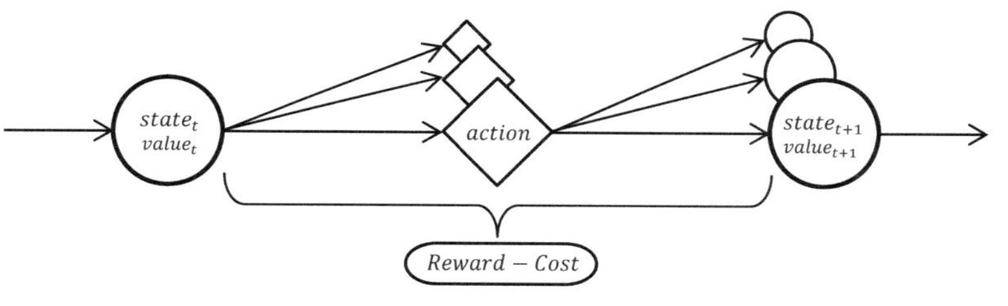

Figure 1. Markov Decision Process depiction

According to the Bellman´s Principle of Optimality, for every MDP can be determined a series of Value Functions, which represents the continuation value of each state. The continuation value associated to a given state is the expected sum of rewards that the optimal strategy would yield from that state until the end of the process (or the expected average reward if the MDP is infinite).

It is easy to see how the value functions of a MDP can be found using a classic backwards DP algorithm. Starting from the final states, a DP algorithm exhaustively calculates the continuation value for a discrete number of states. All these continuation values, collected in a lookup table, constitute later the Value Functions which are accurate but not very compact or easy to calculate. After acquiring the Value Functions, it is simple to find an optimal decision for each state as the one that maximizes the sum of the expected reward and the expected continuation value of the next state or states. However, the problems that a DP algorithm can address by this procedure are restricted by the size of their state spaces.

Other forms to represent and/or approximate the Value Functions can then be proposed. These approaches should not require exhaustive calculation of every state. The Value Functions can be interpolated between computed states. Approximate Dynamic Programming algorithms are built on this cornerstone: approximation of the Value Functions in the state space domain. The estimation methods can be linear regressions, artificial neural networks, etc. Several authors make detailed analysis of MDPs and the use of ADP and DP algorithms to solve them [2],[3].

For approximating and updating the Value Functions, the proposed algorithm uses linear regression on Gaussian radial basis functions jointly with Monte Carlo simulations to consider randomness. An interior-point optimization algorithm is implemented to make decisions.

The ADP algorithm starts with a series of approximations of the value functions, usually constant. Then taking a Monte Carlo sample, a simulation of the system is conducted. At each decision stage, the algorithm makes a decision that is optimal regarding to the current state and the available approximations. Finally, after each Monte Carlo simulation, decisions and outcomes are used to refine the estimation of the Value Functions and complementary Risk Functions, denoted by V_t and R_t respectively. The process continues iteratively, until a certain termination criterion is fulfilled. A simple diagram of this approach is illustrated in Figure 2.

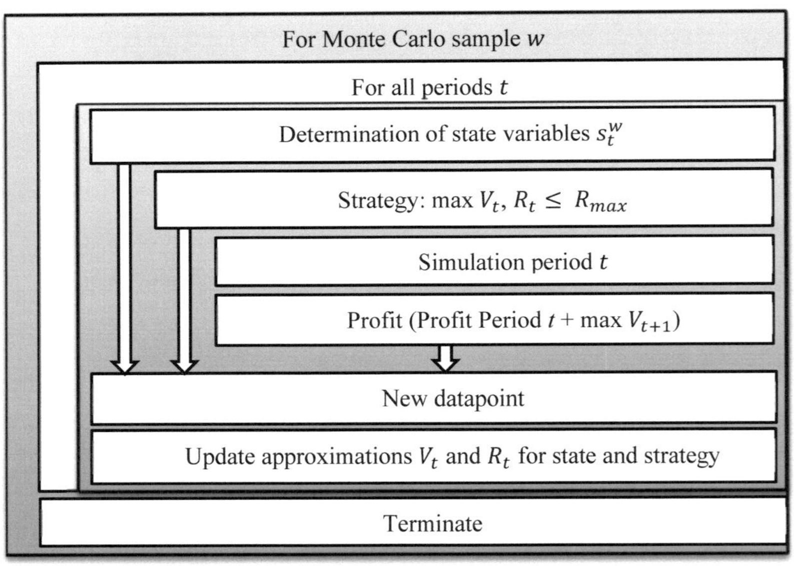

Figure 2. Simulation and estimation update as performed by the proposed ADP algorithm

2.1. Regression technique

An approximation of a system involves finding a function that can predict the output values for a given set of inputs. In this case, the inputs are the state variables and the decision and the outputs are the continuation value and risk.

To approximate the value and risk functions one could use several methods such as linear regressions, artificial neural networks, splines, etc. In the context of electricity trading, it will be discussed how to use linear regression with radial basis functions. The same principles and techniques apply to any regression based on linear parameters.

The main feature of the algorithm is that the approximation is used to make decisions while collecting new data to further improve the currently available approximation. Thus, it is necessary that the regressed function fitting the data can be readily updated as new data is simulated, making the improved approximations immediately available for the next simulation. To fulfill this requirement, a recursive regression technique can be used.

Let be a set of inputs X and a set of collected data from outputs Y

$$\mathbf{X} = \begin{bmatrix} x_{1,1} & x_{1,2} & \cdots & x_{1,g} \\ x_{2,1} & x_{2,2} & \cdots & x_{2,g} \\ \vdots & \vdots & \ddots & \vdots \\ x_{n,1} & x_{n,2} & \cdots & x_{n,g} \end{bmatrix} = \begin{bmatrix} \mathbf{x}_1 \\ \mathbf{x}_2 \\ \vdots \\ \mathbf{x}_n \end{bmatrix}, \mathbf{Y} = \begin{bmatrix} y_{1,1} & y_{1,2} & \cdots & y_{1,h} \\ y_{2,1} & y_{2,2} & \cdots & y_{2,h} \\ \vdots & \vdots & \ddots & \vdots \\ y_{n,1} & y_{n,2} & \cdots & y_{n,h} \end{bmatrix} = \begin{bmatrix} \mathbf{y}_1 \\ \mathbf{y}_2 \\ \vdots \\ \mathbf{y}_n \end{bmatrix} \tag{1}$$

The matrix of inputs X has n data points and g dimensions and the matrix of outputs Y has n data points and h dimensions. Then, the approximation function is:

$$\hat{Y} = \sum_{j=1}^{k} \varphi_j(X) \cdot \theta^j = V(X) = \begin{bmatrix} \hat{y}_{1,1} & \hat{y}_{1,2} & \cdots & \hat{y}_{1,h} \\ \hat{y}_{2,1} & \hat{y}_{2,2} & \cdots & \hat{y}_{2,h} \\ \vdots & \vdots & \ddots & \vdots \\ \hat{y}_{n,1} & \hat{y}_{n,2} & \cdots & \hat{y}_{n,h} \end{bmatrix} = \begin{bmatrix} \hat{y}_1 \\ \hat{y}_2 \\ \vdots \\ \hat{y}_n \end{bmatrix} = \varphi(X) \cdot \theta \tag{2}$$

$$\boldsymbol{\varphi}(\mathbf{X}) = \begin{bmatrix} \varphi_1(\mathbf{x}_1) & \varphi_2(\mathbf{x}_1) & \cdots & \varphi_k(\mathbf{x}_1) \\ \varphi_1(\mathbf{x}_2) & \varphi_2(\mathbf{x}_2) & \cdots & \varphi_k(\mathbf{x}_2) \\ \vdots & \vdots & \ddots & \vdots \\ \varphi_1(\mathbf{x}_n) & \varphi_2(\mathbf{x}_n) & \cdots & \varphi_k(\mathbf{x}_n) \end{bmatrix} = \begin{bmatrix} \mathbf{z}_1 \\ \mathbf{z}_2 \\ \vdots \\ \mathbf{z}_n \end{bmatrix} = \mathbf{Z},\ \boldsymbol{\theta} = \begin{bmatrix} \theta^1 \\ \theta^2 \\ \vdots \\ \theta^k \end{bmatrix} \tag{3}$$

Where the kernel functions $\varphi(X)$ transform the input variables X into a k-dimensional space and are in general non-linear. Here, it is important to notice that despite the fact the φ functions are nonlinear, the approximation is still linear with respect the parameters θ. Therefore, the regression parameters can easily be found by solving a linear system of equations.

The parameter vector θ that minimizes the mean quadratic error MQE of the estimated outputs is:

$$\boldsymbol{\theta} = \left[\mathbf{Z}^{\mathrm{T}}\mathbf{Z}\right]^{-1}\mathbf{Z}^{\mathrm{T}}\mathbf{Y} = \mathbf{A}\,\mathbf{Z}^{\mathrm{T}}\mathbf{Y} \tag{4}$$

$$\mathbf{A} = \left[\mathbf{Z}^{\mathrm{T}}\mathbf{Z}\right]^{-1} \tag{5}$$

The dimension of matrices X, Y and Z increase as new points are simulated, and thus the estimated parameters change as new data is collected. On the other hand, the algorithm needs the approximation to get new data. One approach could be calculating the new parameters after a number of simulations with equation (4) but it soon reveals impractical as an increasing amount of data need to be stored. Instead, by storing only the matrix A and the parameter vector θ, the matrix itself can recursively be updated jointly with the parameters, following the process described in [2]:

In the i-th iteration, a new data point x_i, y_i is simulated and the matrices X and Y become:

$$\mathbf{X}_i = \begin{bmatrix} \mathbf{X}_{i-1} \\ \mathbf{x}_i \end{bmatrix}, \mathbf{Y}_i = \begin{bmatrix} \mathbf{Y}_{i-1} \\ \mathbf{y}_i \end{bmatrix} \tag{6}$$

Also the matrix Z becomes:

$$\mathbf{Z}_i = \begin{bmatrix} \varphi(\mathbf{X}_{i-1}) \\ \varphi(\mathbf{x}_i) \end{bmatrix} = \begin{bmatrix} \mathbf{Z}_{i-1} \\ \mathbf{z}_i \end{bmatrix} \tag{7}$$

Replacing these values in (4):

$$\boldsymbol{\theta}_i = \left[\mathbf{Z}_i^{\mathrm{T}}\mathbf{Z}_i\right]^{-1} \mathbf{Z}_i^{\mathrm{T}}\mathbf{Y}_i = \mathbf{A}_i \mathbf{Z}_i^{\mathrm{T}}\mathbf{Y}_i \tag{8}$$

$$\boldsymbol{\theta}_i = \left[\begin{bmatrix} \mathbf{Z}_{i-1}^{\mathrm{T}} & \mathbf{z}_i^{\mathrm{T}} \end{bmatrix} \begin{bmatrix} \mathbf{Z}_{i-1} \\ \mathbf{z}^i \end{bmatrix}\right]^{-1} \begin{bmatrix} \mathbf{Z}_{i-1}^{\mathrm{T}} & \mathbf{z}_i^{\mathrm{T}} \end{bmatrix} \begin{bmatrix} \mathbf{Y}_{i-1} \\ \mathbf{y}_i \end{bmatrix} \tag{9}$$

$$\boldsymbol{\theta}_i = \left[\mathbf{Z}_{i-1}^{\mathrm{T}}\mathbf{Z}_{i-1} + \mathbf{z}_i^{\mathrm{T}}\mathbf{z}_i\right]^{-1} \left[\mathbf{Z}_{i-1}^{\mathrm{T}}\mathbf{Y}_{i-1} + \mathbf{z}_i^{\mathrm{T}}\mathbf{y}_i\right] \tag{10}$$

Using the Sherman-Morrison formula:

$$\left[\mathbf{L} + \boldsymbol{u}^{\mathrm{T}}\boldsymbol{u}\right]^{-1} = \mathbf{L}^{-1} - \frac{\mathbf{L}^{-1}\, \boldsymbol{u}^{\mathrm{T}}\boldsymbol{u}\mathbf{L}^{-1}}{1 + \boldsymbol{u}\mathbf{L}^{-1}\, \boldsymbol{u}^{\mathrm{T}}} \tag{11}$$

Where L is a $k \times k$ invertible matrix and u is a k-dimensional row vector:

$$\boldsymbol{\theta}_i = \left(\left[\mathbf{Z}_{i-1}{}^{\mathrm{T}} \mathbf{Z}_{i-1} \right]^{-1} - \frac{\left[\mathbf{Z}_{i-1}{}^{\mathrm{T}} \mathbf{Z}_{i-1} \right]^{-1} \mathbf{z}_i{}^{\mathrm{T}} \mathbf{z}_i \left[\mathbf{Z}_{i-1}{}^{\mathrm{T}} \mathbf{Z}_{i-1} \right]^{-1}}{1 + \mathbf{z}_i \left[\mathbf{Z}_{i-1}{}^{\mathrm{T}} \mathbf{Z}_{i-1} \right]^{-1} \mathbf{z}_i{}^{\mathrm{T}}} \right) \left(\mathbf{Z}_{i-1}{}^{\mathrm{T}} \mathbf{Y}_{i-1} + \mathbf{z}_i{}^{\mathrm{T}} \mathbf{y}_i \right) \tag{12}$$

Replacing $[Z_{i-1}{}^{\mathrm{T}} Z_{i-1}]^{-1} = A_{i-1}$

$$\boldsymbol{\theta}_i = \left(\mathbf{A}_{i-1} - \frac{\mathbf{A}_{i-1} \mathbf{z}_i{}^{\mathrm{T}} \mathbf{z}_i \ \mathbf{A}_{i-1}}{1 + \mathbf{z}_i \mathbf{A}_{i-1} \mathbf{z}_i{}^{\mathrm{T}}} \right) \left(\mathbf{Z}_{i-1}{}^{\mathrm{T}} \mathbf{Y}_{i-1} + \mathbf{z}_i{}^{\mathrm{T}} \mathbf{y}_i \right) \tag{13}$$

Naming $\rho_i = \frac{1}{1 + z_i \ A_{i-1} \ z_i{}^{\mathrm{T}}}$ and distributing:

$$\boldsymbol{\theta}_i = \mathbf{A}_{i-1} \mathbf{Z}_{i-1}{}^{\mathrm{T}} \mathbf{Y}_{i-1} + \mathbf{A}_{i-1} \mathbf{z}_i{}^{\mathrm{T}} \mathbf{y}_i - \rho_i \mathbf{A}_{i-1} \mathbf{z}_i{}^{\mathrm{T}} \mathbf{z}_i \ \mathbf{A}_{i-1} \mathbf{Z}_{i-1}{}^{\mathrm{T}} \mathbf{Y}_{i-1} - \rho_i \mathbf{A}_{i-1} \mathbf{z}_i{}^{\mathrm{T}} \mathbf{z}_i \ \mathbf{A}_{i-1} \mathbf{z}_i{}^{\mathrm{T}} \mathbf{y}_i \tag{14}$$

Now, simplifying $A_{i-1} Z_{i-1}{}^{\mathrm{T}} Y_{i-1} = \theta_{i-1}$

$$\boldsymbol{\theta}_i = \boldsymbol{\theta}_{i-1} - \rho_i \mathbf{A}_{i-1} \mathbf{z}_i{}^{\mathrm{T}} \mathbf{z}_i \ \boldsymbol{\theta}_{i-1} - \rho_i \mathbf{A}_{i-1} \mathbf{z}_i{}^{\mathrm{T}} \mathbf{z}_i \ \mathbf{A}_{i-1} \mathbf{z}_i{}^{\mathrm{T}} \mathbf{y}_i + \mathbf{A}_{i-1} \mathbf{z}_i{}^{\mathrm{T}} \mathbf{y}_i \tag{15}$$

Taking common factor $-\rho_i \ A_{i-1} z_i{}^{\mathrm{T}}$ of the last 3 terms:

$$\boldsymbol{\theta}_i = \boldsymbol{\theta}_{i-1} - \rho_i \mathbf{A}_{i-1} \mathbf{z}_i{}^{\mathrm{T}} \left[\mathbf{z}_i \ \boldsymbol{\theta}_{i-1} + \mathbf{z}_i \ \mathbf{A}_{i-1} \mathbf{z}_i{}^{\mathrm{T}} \mathbf{y}_i - \frac{1}{\rho_i} \mathbf{y}_i \right] \tag{16}$$

Replacing $\frac{1}{\rho_i} = 1 + z_i \ A_{i-1} \ z_i{}^{\mathrm{T}}$:

$$\boldsymbol{\theta}_i = \boldsymbol{\theta}_{i-1} - \rho_i \mathbf{A}_{i-1} \mathbf{z}_i{}^{\mathrm{T}} \left[\mathbf{z}_i \ \boldsymbol{\theta}_{i-1} + \mathbf{z}_i \ \mathbf{A}_{i-1} \mathbf{z}_i{}^{\mathrm{T}} \mathbf{y}_i - \left(1 + \mathbf{z}_i \mathbf{A}_{i-1} \mathbf{z}_i{}^{\mathrm{T}} \right) \mathbf{y}_i \right] \tag{17}$$

$$\boldsymbol{\theta}_i = \boldsymbol{\theta}_{i-1} - \rho_i \mathbf{A}_{i-1} \mathbf{z}_i{}^{\mathrm{T}} \left[\mathbf{z}_i \ \boldsymbol{\theta}_{i-1} + \mathbf{z}_i \ \mathbf{A}_{i-1} \mathbf{z}_i{}^{\mathrm{T}} \mathbf{y}_i - \mathbf{y}_i - \mathbf{z}_i \mathbf{A}_{i-1} \mathbf{z}_i{}^{\mathrm{T}} \mathbf{y}_i \right] \tag{18}$$

Finally, canceling out we get a formula to update the parameter vector:

$$\boldsymbol{\theta}_i = \boldsymbol{\theta}_{i-1} - \rho_i \mathbf{A}_{i-1} \mathbf{z}_i{}^{\mathrm{T}} \left[\mathbf{z}_i \ \boldsymbol{\theta}_{i-1} - \mathbf{y}_i \right] \tag{19}$$

Note that in order to update the approximations, it is needed to store and update only the values of the matrix A and the parameter vector θ. The complete set of update formulas is then:

$$\rho_i = \frac{1}{1 + \mathbf{z}_i \, \mathbf{A}_{i-1} \, \mathbf{z}_i^{\mathrm{T}}} \tag{20}$$

$$\boldsymbol{\theta}_i = \boldsymbol{\theta}_{i-1} - \rho_i \, \mathbf{A}_{i-1} \mathbf{z}_i^{\mathrm{T}} \left[\mathbf{z}_i \, \boldsymbol{\theta}_{i-1} - \mathbf{y}_i \right] \tag{21}$$

$$\mathbf{A}_i = \mathbf{A}_{i-1} - \rho_i \, \mathbf{A}_{i-1} \mathbf{z}_i^{\mathrm{T}} \mathbf{z}_i \, \mathbf{A}_{i-1} \tag{22}$$

By using the equations (20), (21) and (22), it is possible to update a regression while using it to take decisions within the ADP algorithm without explicitly storing the entire dataset and performing the matrix inversion after each Monte Carlo simulation. However, an important issue arises at the start of the recursive process, namely the starting values for both the parameter vector θ and the matrix A. One possibility is to carry first a number of simulations with random decisions, and then use the data collected to calculate the starting values for θ and A. Another option is to use a diagonal matrix:

$$\mathbf{A}_i = \begin{bmatrix} a_{1,1} & 0 & \dots & 0 \\ 0 & a_{2,2} & \dots & 0 \\ \vdots & \vdots & \ddots & \vdots \\ 0 & 0 & \dots & a_{k,k} \end{bmatrix} \tag{23}$$

This option is usually simpler despite the fact that setting the absolute values of the elements requires care and it is not always clear in the literature. To set suitable values, a closer look at the regression formulas for matrix A is needed:

$$\mathbf{A}_i = \left[\mathbf{Z}_i^{\mathrm{T}} \mathbf{Z}_i \right]^{-1} = \left[\mathbf{Z}_{i-1}^{\mathrm{T}} \mathbf{Z}_{i-1} + \mathbf{z}_i^{\mathrm{T}} \mathbf{z}_i \right]^{-1} \tag{24}$$

$$\mathbf{A}_i = \left[\mathbf{Z}_i^{\mathrm{T}} \mathbf{Z}_i \right]^{-1} = \left[\mathbf{Z}_{i-2}^{\mathrm{T}} \mathbf{Z}_{i-2} + \mathbf{z}_{i-1}^{\mathrm{T}} \mathbf{z}_{i-1} + \mathbf{z}_i^{\mathrm{T}} \mathbf{z}_i \right]^{-1} \tag{25}$$

$$\mathbf{A}_i = \left[\mathbf{Z}_i^{\mathrm{T}} \mathbf{Z}_i \right]^{-1} = \left[\sum_{p=1}^{i} \mathbf{z}_p^{\mathrm{T}} \mathbf{z}_p \right]^{-1} \tag{26}$$

$$\mathbf{A}_i = \left[\sum_{p=1}^{i} \begin{bmatrix} \left(z_{p,1}\right)^2 & z_{p,1}z_{p,2} & \cdots & z_{p,1}z_{p,k} \\ z_{p,1}z_{p,2} & \left(z_{p,2}\right)^2 & \cdots & z_{p,2}z_{p,k} \\ \vdots & \vdots & \ddots & \vdots \\ z_{p,1}z_{p,k} & z_{p,2}z_{p,k} & \cdots & \left(z_{p,k}\right)^2 \end{bmatrix} \right]^{-1} \tag{27}$$

The matrix A_i is the inverse of a matrix sum whose diagonal elements are positive and increasing with i. That means that the absolute value of the diagonal elements of A would in general decrease as i increases. The relation factor between the starting value of all the elements $a_{j,j}$ and the corresponding expected value of $(1/z_{p,j})^2$ controls how new values modify the parameter vector $\boldsymbol{\theta}$.

$$a_{j,j} = \delta \cdot \mathsf{E}\left[1/\left(z_{p,j}\right)^2\right] \tag{28}$$

Setting starting diagonal elements with a small δ value implies implies a large number of iterations i. This would mean that the starting parameter vector represents already a large amount of "fictional" data. In that case, any new real data sample will have little impact on the parameter vector, and the convergence rate of the approximation might be significantly slowed. On the other hand, setting initial diagonal values calculated with large δ enables the algorithm to large modifications on the parameter vector as new data is collected, which could cause instability of the parameter vector and in the decision computed, again reducing the overall convergence rate. In practice, the factor δ depends on the problem and it must be assessed accordingly. From extensive experimentation, values of δ between 1 and 1000 tend to work most of the time.

For initializing the parameter vector $\boldsymbol{\theta}_0$, it is usual to set a constant vector. Values obtained from the formula $\hat{Y}_s = \varphi(\mathsf{E}(X)) \cdot \boldsymbol{\theta}_0$ produce coherent results. It is important to notice that the starting values of the parameter vector are important in cases where δ is small, while for a large δ the starting vector value is lost in the first iterations.

2.2. Kernel functions

The approximations are highly influenced by the type and parameterization of the kernel functions. These functions transform the original inputs x_m, $1 \le m \le g$ into a k-dimensional space where a linear correlation between outputs and transformed inputs is more accurate.

There are several kernel functions, such as polynomial, trigonometric, logarithmic, radial basis, etc. The algorithm proposed in this work uses Gaussian radial basis functions, whose general formula is:

$$\varphi_j(x) = e^{-(\alpha \| x - c_j \|)^2} \tag{29}$$

where c_j is 1xg vector called a center or centroid, with g being the number of input variables.

The functions measure the distance of each variable in the input state space to k centers and then transform each distance using a Gaussian function. The Euclidean norm is typically used to measure the distance. Nevertheless, any other norm can be considered.

The number and relative position of centers c_j are important. The number of centers determines the dimension of the matrices in the linear regression, and also the smoothing and overfitting properties of the approximation. There must be enough centers and they must be placed to "cover" the entire state space, avoiding blanks in regions of interest while keeping the number of centers to a minimum. A random setting of the centers in the state space can serve in some cases. However, it is usually more appropriate to place the centers using techniques such as Latin Hyper Cube Sampling, as it is done in the implemented algorithm.

2.3. ADP and linear regressions in the context of distributed computing

In the context of ADP, a large amount of the calculation is attributed to the Monte Carlo simulations. Consequently, distributed computing techniques can be exploited to dramatically improve and speed up the optimization. The optimal point between thread opening and result gathering basically depends on the hardware system, network latency and simulation times. With time-consuming simulations, it is often useful to parallelize the program after each approximation update, leaving several threads gather more data in parallel before any update. But if the simulation time is short, the overhead times to distribute the calculation are usually larger than the speed up achieved through parallel simulation. In such cases, another approach should be used. A practical approach is to run parallel ADP algorithms for the same problem in several threads in order to combine and synchronize all approximations at regular time intervals. The combination of the results of two or more independent threads is based on the same mathematical basis as the recursive update, provided that the non-linear parametric functions are identical for all threads.

As shown in equation (4), the optimal parameter vector is:

$$\boldsymbol{\theta} = \left[\mathbf{Z}^{\mathrm{T}}\mathbf{Z}\right]^{-1}\mathbf{Z}^{\mathrm{T}}\mathbf{Y} = \left[\sum_{i=1}^{n}\mathbf{z}_i^{\mathrm{T}}\mathbf{z}_i\right]^{-1} \cdot \left[\sum_{i=1}^{n}\mathbf{z}_i^{\mathrm{T}}\mathbf{y}_i\right] \tag{30}$$

$$\boldsymbol{\theta} = \left[\mathbf{B}\right]^{-1}\mathbf{C},\ \mathbf{B} = \sum_{i=1}^{n}\mathbf{z}_i^{\mathrm{T}}\mathbf{z}_i,\ \mathbf{C} = \sum_{i=1}^{n}\mathbf{z}_i^{\mathrm{T}}\mathbf{y}_i \tag{31}$$

where each z_i and y_i represent the transformed inputs and the outputs respectively of the i-th simulation of a total of n data points. As i increases, the matrices $\boldsymbol{B}$ and $\boldsymbol{C}$ summarize all the regression data collected. The data simulated by several threads of ADP can be easily combined using these matrices. As parallel ADP threads a and b gather different sets of data, the parameter vectors that they approximate and update are different

$$\boldsymbol{\theta}^a = \left[\mathbf{B}^a\right]^{-1}\mathbf{C}^a = \left[\sum_{i=1}^{v}\mathbf{z}_i^{\mathrm{T}}\mathbf{z}_i\right]^{-1}\cdot\left[\sum_{i=1}^{v}\mathbf{z}_i^{\mathrm{T}}\mathbf{y}_i\right] \tag{32}$$

$$\boldsymbol{\theta}^b = \left[\mathbf{B}^b\right]^{-1}\mathbf{C}^b = \left[\sum_{i=v}^{n}\mathbf{z}_i^{\mathrm{T}}\mathbf{z}_i\right]^{-1}\cdot\left[\sum_{i=v}^{n}\mathbf{z}_i^{\mathrm{T}}\mathbf{y}_i\right] \tag{33}$$

However, saving the matrices $\boldsymbol{B}$ and $\boldsymbol{C}$ allows to recombine all the data later as

$$\boldsymbol{\theta} = \left[\sum_{i=1}^{v}\mathbf{z}_i^{\mathrm{T}}\mathbf{z}_i + \sum_{i=v}^{n}\mathbf{z}_i^{\mathrm{T}}\mathbf{z}_i\right]^{-1}\cdot\left[\sum_{i=1}^{v}\mathbf{z}_i^{\mathrm{T}}\mathbf{y}_i + \sum_{i=v}^{n}\mathbf{z}_i^{\mathrm{T}}\mathbf{y}_i\right] \tag{34}$$

$$\boldsymbol{\theta} = \left[\mathbf{B}^a + \mathbf{B}^b\right]^{-1}\cdot\left[\mathbf{C}^a + \mathbf{C}^b\right] \tag{35}$$

This result can be generalized to any number of parallel threads.

In general, data gathered from several parallel algorithms can be recombined by updating and saving the matrices $\boldsymbol{B}$ and $\boldsymbol{C}$ after each simulation i of each thread j. The update formulas are:

$$\mathbf{B}_i^j = \mathbf{B}_{i\text{-}1}^j + \mathbf{z}_i^{\mathrm{T}}\mathbf{z}_i \tag{36}$$

$$\mathbf{C}_i^j = \mathbf{C}_{i-1}^j + \mathbf{z}_i^{\mathrm{T}}\mathbf{y}_i \tag{37}$$

After a fixed simulation time, they can be synchronized and recombined, so all the threads share the same data gathered:

$$\mathbf{B} = \sum_{j=1}^{th}\mathbf{B}^j \tag{38}$$

$$\mathbf{C} = \sum_{j=1}^{th}\mathbf{C}^j \tag{39}$$

Finally, each thread can restart the simulation process using as starting values for θ and A

$$\mathbf{A} = \left[\mathbf{B}\right]^{-1} \tag{40}$$

$$\boldsymbol{\theta} = \left[\mathbf{B}\right]^{-1}\mathbf{C} \tag{41}$$

The final approximation is not as good as the one carried out by a single thread for the same amount of simulations. This is due to the fact that the decisions taken by the single thread algorithm have always all the information gathered up to the decision point while the multithread algorithm lacks the information gathered by other threads since the last synchronization. Nevertheless, with the correct choice of synchronization and simulation cycles, the overall optimization process can run much faster.

3. Electricity trading

3.1. Electricity markets

Since about two decades, the power industry had undergone a major restructuring in many countries. The former vertically-owned and centrally-planned electricity utilities have been unbundled in separate and independent business segments: the generation, transmission and distribution sectors. Unlike the latter two segments, which have remained as natural monopolies under regulation, power generation is now a business subject to competition in the open marketplace [5].

Electricity is a commodity with some very distinctive features. First, modern societies are exceedingly dependent on a continuous delivery of electrical power, placing a very high value to supply reliability. Because electrical energy cannot be economically stored in considerable amounts, production and consumption must be continuously in perfect balance. In addition, power demand is nearly price irresponsive (inelastic) in the short-run. Therefore, power prices often escalate to very high quotes (price spikes) when supply/demand conditions are tight. Most of these circumstances are short-lived, e.g. equipment outages, transmission congestion climatic events, etc., and price rapidly reverse to normal levels [6].

The exceptionally high volatility of electricity prices imposes high financial risks when trading electricity and forces generation companies to make decisions and commitments under high uncertainty. Thus, stochastic modeling and optimal decision making under uncertainty are key tasks in modern power trading and power risk management [7].

3.2. Trading power in spot markets

Currently, electricity is traded forward on bilateral negotiations and on centrally-run electronic platforms like power exchanges. Electricity can be traded in anticipation from one to three years (mostly OTC) to several months ahead. Shorter term forward markets negotiating electricity with delivery horizon of weeks, or even one day in advance in the so-called day-ahead markets are quite common. Market liquidity is typically higher as the contracting horizon shortens.

Because of technical requirements of real time system balance, spot power markets are always centralized and run by entities in charge of the physical operation of the system to keep reliability and security. Equilibrium spot prices are computed each 5 min, 30 min or in hourly basis according to the realized power demand and the last bid accepted for keeping the system balance in real time. Rigorously, locational marginal prices are set in order to account for transmission constraints. Therefore, spot prices reflect the actual conditions of the system at the time of delivery.

Though spot prices are subject to high uncertainty, liquidity of this market is warranted as the generator can always sell its production at real time prices. For this reason, the spot market is regarded a last-resort market. One advantage of participating only in this market is that unavailability of the generating unit does not have financial consequences for the generator other than the opportunity cost of the lost production.

Let $p_S(t)$ the prevailing price at the t-th time interval in the spot market, $P_S(t)$ the power delivered by the generating unit and $P_{\max}$ its generating capacity. Under the hypothesis the generator is price-taker and its marginal costs of generation, denoted by $MC(t)$, are constant with the rate of production, the optimal operating policy is:

$$P_S(t) = \begin{cases} 0 & \text{if } p_S(t) < MC(t) \\ P_{max} & \text{if } p_S(t) > MC(t) \end{cases} \tag{42}$$

The operating profit $B_S(t)$ the generator obtains in the spot market by implementing this optimal production policy can be written as:

$$B_s(t) = \max\left[p_s(t)P_{max} - MC(t)P_{max}, 0\right] = \max\left[\left(p_s(t) - MC(t)\right)P_{max}, 0\right] \tag{43}$$

This equation shows the operating flexibility of generator to alter its output in response to the spot price in order to avoid operating losses if prevailing spot prices drop below marginal costs.

Figure 3 depicts the discontinuous nature of the profit function $B_S(t)$ when participating in the spot market, i.e. $B_S(t) \geq 0$ for all prices. Indeed, this profit function can be assimilated to a call option with strike price MC. Figure 3 also schematically illustrates the probability density function (pdf) of the spot price of electricity $f(p_S)$. This function is typically highly right-skewed and presents strong leptokurtosis.

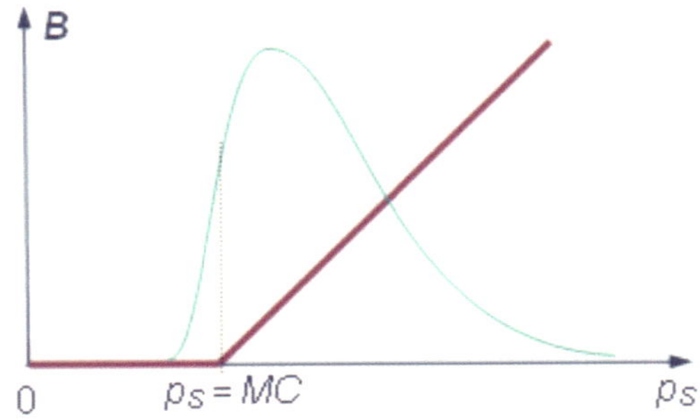

Figure 3. Profit function of selling power in the spot market

The expected value of the profit in the spot market per unit of generating capacity $b_S(t)$ under the optimal operating policy is described by the following equation:

$$\mathrm{E}\left[b_S\right]=\int_0^{MC}\left[0\cdot p_S f(p_S)\right]dp_S+\int_{MC}^{\infty}\left[\left(p_S-MC\right)f(p_S)\right]dp_S \tag{44}$$

$$\mathrm{E}\left[b_S\right]=\int_{MC}^{\infty}\left[\left(p_S-MC\right)f(p_S)\right]dp_S\geq \mathrm{E}\left[p_S\right]-MC \tag{45}$$

It is noteworthy to observe that the first term in equation is the probability of obtaining a zero profit in the spot market. Note also that $\mathrm{E}[p_S]=MC$ only if $MC=0$. By selling the production in the spot market, the generator never incurs in operating losses, i.e. $\Pr(b_S<0)=0$ as it can immediately stop production if $p_S<MC$.

We consider now the more general case where generating units are unavailable, either for planned or unplanned reasons, during a fraction of the time. Let p the failure probability and $q=1-p$ the probability of the unit being available, provided the failure and operating states are the only two mutually exclusive states in which the generator resides. We further assume that the price level and the state of the generator are statistically independent. Under these considerations, the generator cannot always capture de spread p_S-MC and thus the probability of obtaining a positive profit will decrease accordingly. The expected operating profit under these conditions is then:

$$\mathrm{E}\left[b_S\right]=q\int_{MC}^{\infty}\left[\left(p_S-MC\right)f(p_S)\right]dp_S \tag{46}$$

The probability of having zero profit $\beta_0=\Pr(b_S=0)$ is given by:

$$\beta_0 = p\int_0^{\infty} f(p_S)dp_S + q\int_0^{MC} f(p_S)dp_S = p + q\int_0^{MC} f(p_S)dp_S \tag{47}$$

The resulting probability density function of the hourly operating profit is illustrated in Figure 4. Despite the high variance of the profits, the function clearly shows that the generator cannot lose money when participating in the spot market, even if the unit is technically unavailable.

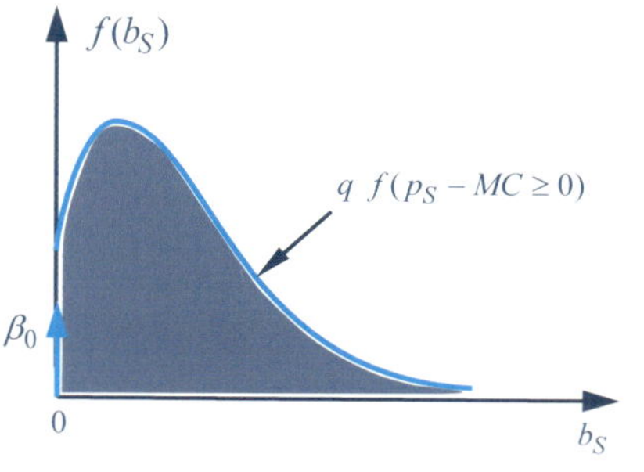

Figure 4. Probability density function of the operating profit in the spot market of electricity

3.3. Trading electricity in forward markets

Given the dramatic volatility of real time electricity prices, a major activity of power trading is structuring hedging strategies by means of tradable derivative instruments like future and option contracts [8]. A power company owning a set of generating units may decide either to sell electricity in advance at a fixed price in a forward market, or wait to the time of delivery and receive the spot price. Deciding on committing production forward or being exposed to volatility of real-time power prices has however a drastic impact on risk.

By selling forward its production, the generator may hedge against a sudden decline of electricity spot prices during the delivery horizon, thereby securing an operating margin. This hedging strategy isolates the generator from the price risk. However, the generator in exchange resigns the opportunity of selling electricity in the spot market if high prices happen.

Electricity markets are typically arranged under a two-settlement system. This approach preserves the economy and efficiency of the physical operation of the power system from any financial commitment the market players have entered into in the past. Under the two-settlement scheme, only deviations from contractual obligations are negotiated in the spot market.

The revenue from the forward contracting is given by the volume sold P_F times the price p_F agreed in the forward contract, i.e. $R_F = p_F P_F$. On the other hand, the revenue captured by selling in the spot market is given by $R_S = p_S \Delta P = p_S(P_S - P_F)$. So, the total revenue R_F from forward contracting and delivering power in the spot market is given as:

$$R_T = R_F + R_S = p_F P_F + p_S(P_S - P_F) = P_F(p_F - p_S) + p_S P_S \tag{48}$$

We can observe the utility of forward contracts by inspecting this equation. If generator delivers in the spot market an amount equal to its contractual obligation, i.e. $P_S = P_F$, the total revenue is set equal to $p_F P_F$ irrespective of the fluctuations of the spot price p_S.

At the time of delivery, and assuming the generator is price-taker, the term $P_F(p_F - p_S)$ is fixed and represents the profit of the forward contract against the spot market. Therefore, the profit-maximizing production policy is the same and given by the spot price, irrespective of the contractual obligations. The profit the generator can make by selling electricity in forward markets is given by the expression:

$$B_F = P_F\left(p_F - p_S(t)\right) + \max\left[P_S\left(p_S(t) - MC\right), 0\right] \tag{49}$$

In Figure 5A, a probability density function of the spot price is depicted. In the following, it is assumed that the forward market price is an unbiased estimator of the spot price at the time of delivery. Therefore, the condition $p_F = \mathsf{E}(p_S)$ holds. In the forward contract, the generator makes a profit for unit of capacity $b_F = p_F - MC$, assuming $p_F > MC$. Otherwise, the generator is better by avoiding entering into a forward obligation with negative profit. For realizing this profit, the generator must be able to deliver in the spot market the contracted volume in the exact amount. This profit level is achieved as long as the spot price exceeds the marginal cost, i.e. the probability of making this profit is $\Pr(b_F = p_F - MC) = \Pr(p_S > MC)$.

Graphically, this probability is represented by the dark grey area under the pdf of the spot price (cf. Figure 5A). The generator can make additional profits in the forward contract, $b_F = p_F - p_S > p_F - MC$, each time the spot price drops below the marginal cost, i.e. $p_S < MC$. In fact, the generator is better buying replacement power in the spot market than incurring in fuel costs generating with its own facilities. Figure 5B illustrates the pdf of the profit of a forward contract. When compared with the profit distribution in the spot market (cf. Figure 5), it is easily noticeable the drastic reduction of the profit variance under forward contracting. The forward obligation sets a floor for profits, reducing dramatically dispersion of results and thereby the price risk. In exchange, the generator also foregoes the chance of profiting at times of high power prices in the spot market. The expected profit of a forward contract in terms of the pdf of the spot price $f(p_S)$ can be expressed as:

$$\begin{aligned} \mathrm{E}\left[b_F\right] &= \left(p_F - MC\right)\int_{MC}^{\infty} f(p_S)dp_S + \int_{0}^{MC}\left[\left(p_F - p_S\right)f(p_S)\right]dp_S = \\ &= p_F - MC\int_{MC}^{\infty} f(p_S)\,dp_S - \int_{0}^{MC}\left[p_S f(p_S)\right]dp_S \end{aligned} \tag{50}$$

It can be mathematically demonstrated that under rational expectations and efficiency of forward markets, i.e. the forward price is an unbiased estimator of the spot price $p_F = \mathsf{E}(p_S)$, the condition $E(b_F)=E(b_S)$ holds [1]. This means that for a risk-neutral generator both policies, either selling in the spot market or hedging in forward markets, are entirely equivalent. Nevertheless, for risk-averse players (which is the rule in real market settings), the hedged strategy is clearly preferred as profit expectation remain unaltered while price risk is eliminated.

3.4. Delivery risk in forward contracting

If we consider again unplanned outages of generating units, hedging price risk in the forward markets exposes generating companies to other class of risk, i.e. delivery risk, which also referred as quantity or volume risk. We further examine this important issue. When a generator under a contractual obligation is unable to deliver in the spot market the contracted amount, i.e. $P_S \neq P_F$, the generator is forced to buy replacement power in the spot market at the prevailing price at that time. This may configure a very significant loss if while the generator is down the spot price is considerable higher than its own marginal costs, i.e. $p_S \gg MC$. Under this situation, the generator may be compelled to buy very expensive replacement power to honor the obligation, incurring in a potentially high financial loss. It is interesting to note that if when the unit is unavailable spot prices are lower or equal than the marginal cost, the generator can even make an extra profit $b_F = p_F - p_S > p_F - MC$. The probability density function of the forward position under consideration of positive failure probability and the associated delivery risk is illustrated in Figure 5C.

The expected profit can be computed as the expected value of the contract under the hypothesis of fully reliable unit times the probability of being available:

$$E[b_F] = q\left[(p_F - MC) \int_{MC}^{\infty} f(p_S) dp_S + \int_{0}^{MC} (p_F - p_S) f(p_S) dp_S \right] \tag{51}$$

As for modern units $q \cong 1$, the change in the expected profit due to unit unavailability is typically negligible. However, downside risk increases substantially.

Assuming statistical independence between the unit's failure and the level of spot prices, the probability of incurring in losses is given by:

$$\Pr(b_F < 0) = p \Pr(p_S > p_F) \tag{52}$$

and the conditional expectation on the value of losses can be written as:

$$L = \int_{P_F}^{\infty} (p_F - p_S) f(p_S) dp_S \mid b_F < 0 \tag{53}$$

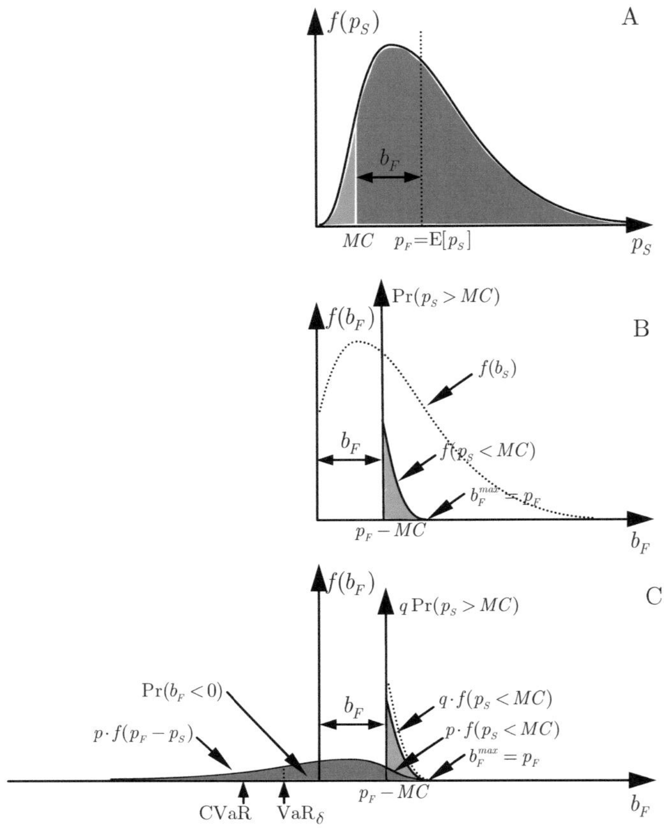

Figure 5. Probability density function of operating profits for a forward contract

From the profit pdf of the forward contract $f(b_F)$, the downside risk metrics, namely the value at risk (VaR) and the conditional value at risk (CVaR) for a δ confidence level can respectively be computed as:

$$\Pr(b_F < -\mathrm{VaR}_\delta) = \int_{-\infty}^{-\mathrm{VaR}} f(b_F)\, db_F = \delta \tag{54}$$

$$\mathrm{CVaR}_\delta = \int_{-\infty}^{-\mathrm{VaR}} b_F f(b_F)\, db_F \tag{55}$$

http://dx.doi.org/10.5772/57466

4. Problem modeling

4.1. Problem formulation

Let consider a small generation portfolio running in an electricity market. The power company owning the generation portfolio wants to determine the best-selling strategy of the energy production, which would maximize the expected profit while financial risk is constrained. Any trading strategy x is defined by the amount of energy to be sold in each different available selling instrument i in the electricity market, for example, the spot market, day-ahead obligations, annual forward contracts, etc.

It is important to notice that the trading strategy sets the amounts of energy committed in every forward instrument, but only estimates the amount of energy to be actually sold in the spot market. Indeed, actual energy production is stochastic and depends on technical availability of generating units and the spread between spot and fuel prices. Suppose that whatever trading strategy is decided now, it could be changed in future decision stages in order to rebalance the portfolio. Composition of the portfolio can be rebalanced only at a cost however, i.e. the transaction costs. The process is then a sequence of balancing decisions determined by the trading strategy, each followed by a stochastic reward according to the position taken in the market. The process is over after a number of periods n.

The optimization of the trading strategy can mathematically be formulated as a stochastic non-linear problem involving the maximization of the expected profit accrued by the generation portfolio across all instruments i and time intervals t:

$$\max_x \left[\mathrm{E}\left[\sum_t \left[\sum_i I^w\left(x_{i,t}\right) - \sum_i T^w\left(x_{i,t}, x_{i,t-1}\right) - C_t^w \right]\right]\right] \tag{56}$$

subject to the following constraints:

$$\sum_{i \neq spot} x_{i,h} + x_{spot,h}^w = E_h^w \ , \ \forall h \tag{57}$$

$$\sum_{i \neq spot} x_{i,t} \leq E_{max,t} \, , \ \forall t \tag{58}$$

$$x_{i,t} \geq 0, \ \forall t, \ i \neq spot \tag{59}$$

$$Risk_W \left[\sum_i I\left(x_{i,t}\right) - \sum_i T\left(x_{i,t}, x_{i,t-1}\right) - C_t^w \right] \leq Risk_{max,t}, \ \forall t \tag{60}$$

where:

E: Expected value operator

w: Monte Carlo sample path

W: Total number of Monte Carlo samples

t: Time period beginning after a balancing decision

h: Hourly time step

$x_{i,h}$: Energy sold and instrument i in hour h

$x_{spot,h}^{w}$: Energy sold in spot market in hour h and Monte Carlo sample w

E_h^w: Energy generated in hour h and Monte Carlo sample w

$E_{max,t}$: Maximal energy that can be generated in period t

$x_{i,t}$: Energy to be sold by instrument i in period t

$I^w(x_{i,t})$: Revenue due to energy the sold by instrument i in period t and Monte Carlo sample w

$$I^w\left(x_{i,t}\right)=\begin{cases} x_{i,t}\cdot p_{F_{i,t}}^{w} ,i=forward \\ \sum_{h\in t} x_{i,h}\cdot p_{S_h}^{w}, \quad i=spot \end{cases} \tag{61}$$

$p_{F_{i,t}}^{w}$: Effective future price of instrument i in period t and Monte Carlo sample w

$p_{S_h}^{w}$: Spot price in hour h and Monte Carlo sample w

C_t^w: Costs of energy in period t and Monte Carlo sample w

$$C_t^w = MC\cdot \sum_{h\in t} E_h^w \tag{62}$$

MC: Constant marginal cost of generation.

$T^w(x_{i,t}, x_{i,t-1})$: Transaction costs due to the change in the amount of instrument i held in the portfolio after the rebalancing decision at the beginning of period t

$$T^w\left(x_{i,t},x_{i,t-1}\right)=\begin{cases} 3\%\cdot p_{F_i}^{w,h0_t}\cdot\left(x_{i,t}-x_{i,t-1}\right) ,i=forward \\ 0 \qquad\qquad\qquad\qquad\qquad , i=spot \end{cases} \tag{63}$$

$p_{F_i}^{w,h0_t}$: Forward price of instrument i and Monte Carlo sample w at the beginning of period t.

The constraint (57) represents the hourly energy balance for all Monte Carlo samples and forces the generator to settle in the spot market differences between the energy sold in forward markets and actual production. Constraints (58) and (59) are introduced to avoid financial positions without physical counterpart, i.e. avoid the generator to take speculative positions by selling in forward markets more energy that the generation portfolio can produce. These constraints may be replaced by capital restrictions as regulations often allow financial trading without physical position. Finally, constraint (60) represents the limit to the financial risk associated to the selling strategy within each period t. In order to limit risk over the horizon time, the selected risk metric must be coherent [9] ensuring subadditivity. There are several downside risk measures that fulfill this requirement among which the Conditional Value at Risk (CVaR) is the most widely used (cf. equation 55).

The equation (61) represents the revenue generated by each forward contract and the revenue (or cost) due to selling (or buying) differences between energy sold in futures and the real generation in the spot market forced by restriction (57). The equation (62) represents the costs of the operating policy, which is independent of the forward obligation for a price taker as demonstrated in Section 3. Therefore, the three equations (57), (61) and (62) calculated for the whole Monte Carlo set represent then the profit calculated by equation (49) of Section 3. Finally the transaction costs in equation (63) are assumed to be the 3% of the total transacted value only for forward contracts, and not existent for the spot market.

4.2. Modeling stochastic spot and forward electricity prices

The problem formulation relies on Monte Carlo simulations to represent uncertainty on the future development of key variables. In addition, stochastic simulations are used to confront the algorithm with mapping scenarios for approximating both, the Value and the Risk functions.

A synthetic ensemble of 2000 annual realizations of hourly power prices in the spot market were generated by means of spectral representation techniques [10]. Forward prices and spot prices are not statistically independent. The forward prices corresponding to each spot price sample are calculated considering both, the expected value of the spot price and the mean value of each spot price time series. To simulate the changes in the forward prices accounting for the correlation to each sample of the spot prices a simple model is introduced.

Under perfect competition and rationality on the expectation formation, the price of a forward product should converge to the mean expected spot prices for the delivery period. This is relatively easy to calculate for the first hour of the simulated time. Assuming that the whole Monte Carlo set of spot prices was simulated taking the same price forecast as the market, the price of any forward should be the mean spot prices for the delivery period:

$$p_{FA\,i}^{\;w,h} = \frac{1}{\left(H_i - \left(h_i^d - 1\right)\right)} \sum_{j=h_i^d}^{H_i} \mathrm{E}\left[p_S{}^j \right] = \frac{1}{W\left(H_i - \left(h_i^d - 1\right)\right)} \sum_{j=h_i^d}^{H_i} \sum_{w=1}^{W} p_S{}^{w,j} \tag{64}$$

$$h_i^d = \begin{cases} H0_i \, if \, h \leq H0_i \\ h \, if \, h > H0_i \end{cases} \tag{65}$$

where:

H_i: delivery period for instrument i

$H0_i$: first hour of delivery period for instrument i

For the first hour, this model represents the expectation of the market for each delivery period. However, this expectation should change according to the values that the spot prices take in each sample and the information gathered by a *virtual* market taking place in each particular sample path. If perfect foresight is assumed, each virtual market could calculate without uncertainty the hourly forward price for its particular spot prices sample path simply as:

$$p_{FB\,i}^{\ w,h} = \frac{1}{H_i - \left(h_i^d - 1\right)} \sum_{j=h_i^d}^{H_i} p_S^{\ w,j} \tag{66}$$

The equation (64) represents a model where any additional information that arrives as the spot price of the particular sample path is different form the forecasted in the first hour is dismissed by the virtual market. Thus, this model is representative of reality only for the first hour of simulation where no additional information could have been gathered by the virtual markets. On the other hand, the equation (66) represents a model where all the additional information is obtained beforehand for each Monte Carlo sample. Likewise, this model is suitable only for the last hour of simulation where all information is already known by the virtual market within each Monte Carlo sample. Finally, the two models can be combined, in order to simulate the forward price dynamics in correlation to the spot prices of each Monte Carlo sample path. In this work, it is assumed that the information gathered by each virtual market grows linearly, augmenting for each hour[1]. Then, equations (64)[2] and (66) are combined by a weighted average:

$$p_{FC\,i}^{\ w,h} = \frac{\left(H - \left(h_i^d - 1\right)\right)}{H_i} \cdot p_{FA\,i}^{\ w,h} + \frac{\left(h_i^d - 1\right)}{H_i} \cdot p_{FB\,i}^{\ w,h} \tag{67}$$

$$\begin{aligned} p_{FC\,i}^{\ w,h} = {} & \frac{\left(H - \left(h_i^d - 1\right)\right)}{H_i} \cdot \frac{1}{W\left(H_i - \left(h_i^d - 1\right)\right)} \sum_{j=h_i^d}^{H_i} \sum_{w=1}^{W} p_S^{\ w,j} + \\ & + \frac{\left(h_i^d - 1\right)}{H_i} \cdot \frac{1}{H_i - \left(h_i^d - 1\right)} \sum_{j=h_i^d}^{H_i} p_S^{\ w,j} \end{aligned} \tag{68}$$

---

1 The assumption that the information is linear with time could be replaced with a more complex information model, such as a function of the cumulated difference between the initial forecasted spot price and the particular spot prices of each Monte Carlo sample path.

2 The prices for the first hour in a real market situation should consider the real market prices.

Finally, a *contango* situation is considered in the forward market. A risk premium of 8% in excess of expected spot prices is considered for the forward prices in the future market to compensate for the volatility risk. This premium reduces linearly during the delivery period of the future and becomes zero at the last hour of delivery.

The model for future prices for each instrument i, hour h and sample w is thus as follows:

$$p_{Fi}^{w,h} = \left(1+\beta\frac{H_i - h_i^d}{H_i}\right)\cdot \left[\frac{\left(H-\left(h_i^d-1\right)\right)}{H_i}\cdot\frac{1}{W\left(H_i-\left(h_i^d-1\right)\right)}\sum_{j=h_i^d}^{H_i}\sum_{w=1}^{W}p_S^{w,j} + \frac{\left(h_i^d-1\right)}{H_i}\cdot\frac{1}{H_i-\left(h_i^d-1\right)}\sum_{j=h_i^d}^{H_i}p_S^{w,j}\right] \tag{69}$$

where β is the risk premium paid in excess to the expected spot price and set β=8%.

4.3. Reliability model of the generation units

Other relevant source of uncertainty considered is the random failure of the generating units. The stochastic model of generator outages is built considering that the unit can reside in four mutually exclusive states: Operation (required), Reserve (not required), Unavailable (required) and Unavailable (not required), as shown in the diagram of space states in Figure 6 [11].

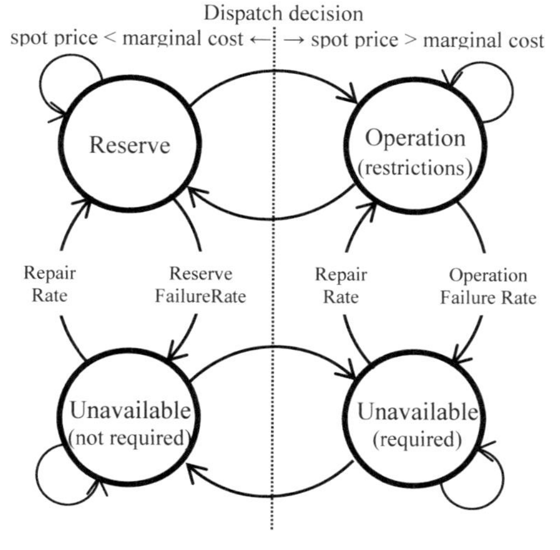

Figure 6. Four-state stochastic model of the generation units

This unit model accounts for the fact that peaking units exhibit higher availability rates. This result is explained by the fact the failure probability is typically very small when the unit is in the stand-by state. A generator is economically called online if its marginal cost of production is below the prevailing spot prices, following the decision model of equation (42). Variable costs of generation are assumed linear with power output, i.e. marginal costs are constant.

The operation-failure cycles of the generating unit are obtained from a chronological Markovian stochastic simulation. For each spot price sample, a time series of power output is synthesized for every generation unit. The hourly power output is simulated o following three steps:

1. Based on failure and repair rates defined by the state the unit resided in the previous hour, a random failure is simulated [12]. If a failure is in place, the output power is set to zero for this particular hour.

2. The dispatch of the unit is simulated, taking into account the marginal cost of generation and the prevailing sample spot price at that time interval. Here perfect foresight of the spot price is assumed in order to decide the dispatch and fulfill the minimal generation times.

3. If dispatched, other unit's technical restrictions are fulfilled, e.g. ramping capabilities.

This chronological stochastic model reproduces with accuracy the dynamics involved in failure and repair cycles of generators, giving the possibility to select different failure rates depending on whether the unit is generating or is in stand-by.

4.4. Risk constraint formulation

As already mentioned before, the financial decision process can be modeled by means of a MDP. Naming profit B^w for each sample the sum of income, cost and transaction cost over all instruments, the objective function (56) becomes:

$$\max_{x}\left[\mathsf{E}\left[\sum_{t}\left[B^{w}\left(x_{t},x_{t-1}\right)\right]\right]\right] \tag{70}$$

Exchanging the order of the expectation operator and the summation, and expanding the summation:

$$\max_{x}\left[\mathsf{E}\left[B_{1}^{w}\left(x_{1},x_{0}\right)\right]+\sum_{t=2}^{n}\mathsf{E}\left[B_{t}^{w}\left(x_{t},x_{t-1}\right)\right]\right] \tag{71}$$

where n is the total number of time periods considered.

At this point and considering the first term depends only on the initial market position x_0 and the first strategy decision x_1, the maximization can be decomposed using the Bellman's Principle of optimality as follows:

$$\max_{x_1}\left[\mathrm{E}\left[B_1^w\left(x_1,x_0\right)\right]+\max_{x_{t=2\ldots n}}\sum_{t=2}^{n}\mathrm{E}\left[B_t^w\left(x_t,x_{t-1}\right)\right]\right] \tag{72}$$

If x is independent of the Monte Carlo sample, the terms inside the summation over all future periods $(t=2\ldots n)$ are simply the expected profit for the period t after a decision $x_{t-1}\rightarrow x_t$. However, the model should take into account that future strategy decisions may be different for each Monte Carlo sample, accounting for adjustments the decision-maker almost certainly would execute to face specific scenarios. Then, the expected profit $\overline{B_t}$ and the decision itself for future decision stages will depend on a set of variables s_t, which represent the variables considered by the decision maker in order to adapt the strategy to a particular situation and the equation (72) becomes:

$$\max_{x_1}\left[\overline{B_1}\left(x_1,x_0,s_1\right)+\max_{x_{t=2\ldots n}}\sum_{t=2}^{n}\overline{B_t}\left(x_t,x_{t-1},s_t\right)\right] \tag{73}$$

Defining the continuation value functions V_t as:

$$V_t\left(x_t,x_{t-1},s_t\right)=\overline{B_t}\left(x_t,x_{t-1},s_t\right)+\max_{x_{t+1}}\left[V_{t+1}\left(x_t,x_{t+1},s_{t+1}\right)\right] \tag{74}$$

The maximization can be solved by a set of recursive maximizations, each one solving only one decision stage:

$$\max_{x_t}\left[\ V_t\left(x_t,x_{t-1},s_t\right)\right] \tag{75}$$

With this model, the optimization can be decomposed in steps and the dynamic nature of a strategy can be accurately replicated. Despite the fact future trading decisions $x_{t=2\ldots n}$ are considered and optimized, the practical product of this procedure is the new optimal rebalanced state $x_{t=1}$ starting from the previous trading position $x_{t=0}$. The further trading positions are only optimal given the current information available and should be reconsidered later. Therefore, each new trading position ($x_{t=2}$, $x_{t=3}$, ..., $x_{t=n}$) should be the product of a similar optimization incorporating the additional market information available immediately before.

The value functions provide the expected continuation value within a state space defined by the state variables, x_t, x_{t-1} and s_t[3]. However, the continuation functions, which are essential to solve the optimization problem, are unknown beforehand. It is here that the ADP approach is introduced to approximate the value and risk functions for the state space.

---

3 There are several other decomposition methods, some of which exclude the decision as state space variable defining the value functions in a post-decision state space. These approaches make the step maximization sometimes harder but present advantages such as a state space of fewer dimensions. See for example [2].

The set of constraints remains the same as they were already defined for each period t. Nevertheless, special considerations should be made to calculate de risk and to fulfill the risk-constrained optimization. As V_t only account for the expected profit, risk functions R_t should also similarly be calculated for the same state space in order to enforce the risk constraint.

By definition, the linear regression will deliver an approximation that minimizes the mean square error on the entire dataset. In a stochastic setting where the same inputs leads to several different outputs, the regression will accurately estimate the expected value for a given set of inputs, provided the sample size and the approximation order are appropriate. This fits perfect for the case of the value function but for the approximation of the risk function some problems arise.

Let suppose that the CVaR is chosen as risk metric. As the algorithm progresses, new data points are collected, i.e. a set of state variables and its corresponding simulated profits for the period. For approximating the CVaR associated to a particular point in the state space, one approach is to select a subset from the dataset whose input variables are "close" to the point. Then the CVaR is calculated first by sorting the profit values and then taking the mean of those below the specified α-quantile. Now, let suppose that a new data point is simulated and an update of the CVaR approximation is needed. To do so, the process described must be repeated, but now including the new data point. This simple approach has large disadvantages: all the data points must be stored and the mean is not easily updated as old data may be excluded or included of the zone below the α-quantile. These drawbacks are caused by the fact that the CVaR is quantile-based. To solve these difficulties, another solution is envisioned. Instead of approximating directly the CVaR, another risk measure is used to approximate the CVaR within the space state. The risk measure used is called Relative Lower Semideviation (RLS) and it is moment based instead. Hence, it can be updated more easily and without needing to store the entire dataset. These types of risk measures are described in detail in [13-16] and for a stochastic profit P_t have the form:

$$RLS(B_t) = -\mathsf{E}[B_t] + a \cdot \sigma_p^{-}[B_t] \tag{76}$$

$$\sigma_p^{-}[B_t] = \left(\mathsf{E}\left[max\left(\mathsf{E}[B_t] - B_t, 0\right)^p\right]\right)^{1/p} \tag{77}$$

where the equation (77) is the negative semideviation of degree p of the stochastic profit P_t.

It can be proven that these moment-based risk measures are coherent if $0 \le a \le 1$ and $p \ge 1$. To approximate a CVaR with a 5%-quantile, in this work the parameters used are $a=1$ and $p=9.5$. To compute the approximation, two linear regressions were used to calculate the expectations on the profit and on the negative deviations, which can be updated using the same method proposed for the value functions. In a Monte Carlo scheme, a large amount of data is needed to compute reliable risk estimations. Therefore, the risk constraint should not

http://dx.doi.org/10.5772/57466

be enforced at the early iterations of the ADP algorithm, simply because the sample size of the dataset is small to get a consistent and statistically converged risk value.

5. Numerical case study

5.1. Algorithm validation

With the objective of validating the results of the proposed ADP algorithm, a first simple exemplary case is considered in which a thermal generator sells energy in the spot market and in a future contract. The results of the ADP algorithm were compared with the results of a conventional DP algorithm for which the space-state was discretized appropriately.

The fractions of energy sold in the spot market and in a quarter future contract were optimized considering that the future can be traded during the delivery period. The optimization determines three decision stages during this period, one at the beginning of each month, consisting on sell or buy energy in the future market based on the previous state. The space-state previous to each decision is defined only by the level of future already sold, in order to keep tractable the DP problem.

5.2. Validation results

The results obtained by solving the problem by means of the ADP and DP algorithms are presented in Figure 7. The plots represent the expected profit and the downside risk measured by the CVaR of the optimal strategy as a function of the initial state, i.e. the energy already committed in the future contract at the initial stage. An excellent agreement between the optimal strategies obtained by ADP and DP is evidenced, validating the proposed approach.

It can be noticed that the expected profit rises as the amount of energy sold forward increases. This is caused by the risk premium paid to the generator in the future market, i.e. the mean future prices are higher than the mean spot prices. Additionally, the transaction costs are not compensated by the risk premium; therefore the best trading strategy is to maintain similar involvement in the future market to the initial level without rebalancing the portfolio. This is illustrated in Figure 8, where the optimal decisions for the first month are practically the same when solving with a conventional DP and an ADP approach.

It is noteworthy to observe in Figure 7 that financial risk lessens when forward contracting in the future market increases. This means that for the conventional generator considered, which present a high availability, the delivery risk is lower than the risk of not being dispatched in the spot market. The behavior of the risk curve is closely related to the unit's failure and reparation rates and to the marginal production costs. Generators with low marginal costs are in the first places of the dispatch merit order, and hence the risk of not being dispatched is low. Moreover, high failure rates imply also a higher delivery risk. Out of these relations arise a broad number of risk curves that differ from one generator to another and suggest that considerable risk mitigation by aggregating different generators in a portfolio is possible.

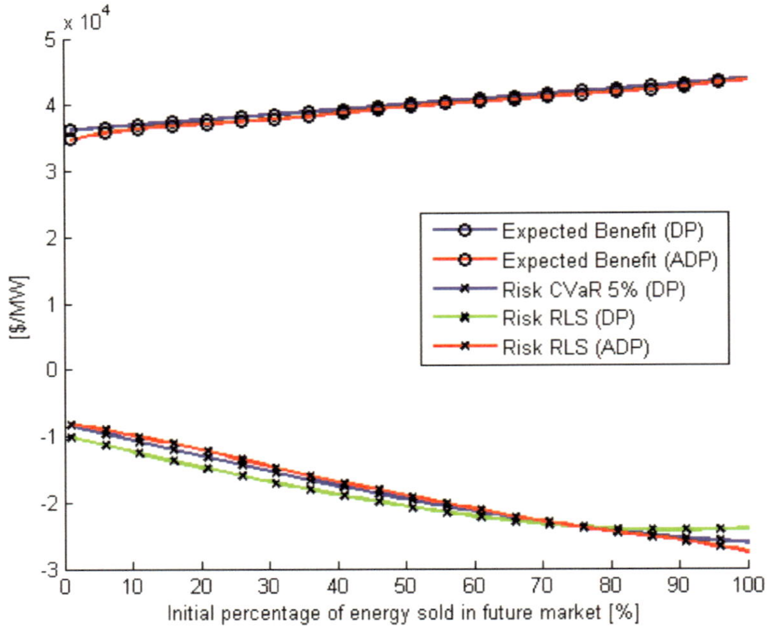

Figure 7. Expected benefit and downside risk of the optimal strategy – Validation case.

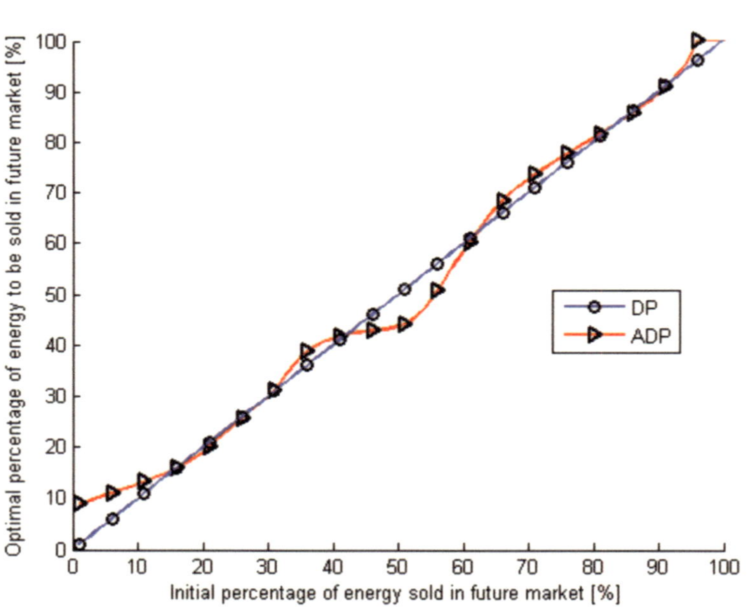

Figure 8. Optimal percentage of energy sold in the future market for the first time period – Validation case

5.3. Optimal policy for portfolio rebalancing

For the base study case, a slightly more complex system was examined. In this case, the generation portfolio comprises five generation units of 2 MW each with a constant marginal generation cost of 50$/MWh, failure rate of 1/950h^{-1}, reserve failure rate of 1/9950 h^{-1} and repair rate of 1/50 h^{-1}. These rates give a failure probability of 5% while the unit is in operation and of 0.5% while in standby.

An annual future, four quarterly futures and the spot market were considered together with 12 monthly trading decision stages for rebalancing the portfolio. The space state for this arrangement is defined by the amount sold forward as well as spot and future prices before each decision point and for each realization. Therefore, the decisions are chosen taking into account additional information about the state of the market for each realization. Even though the simplicity of the example, the additional variables cause a drastic increase in the dimension of the problem and would force an unacceptable coarse discretization of the space-state in order to keep the problem tractable with conventional DP. In this case, the maximum admissible expected loss for the 5%-confidence level is set to $CVaR_{5\%}$=$20000 for each decision stage. Since the risk measure is coherent, i.e. sub-additive, the annual risk is less or equal to $20000•12 periods=$240000. Transaction costs are set to 3% of the dollar amount contracted in the forward market. The optimization problem was solved on a Beowulf cluster comprising 20 multicore Intel i7 2600K 3.4 GHz processors connected by a Gbit LAN. The 160 available computation nodes were fully exploited and the total computation time was 5 h.

In Figure 9, the results of the ADP algorithm for the optimal strategy on the first rebalancing period with a previous trade equal to zero are illustrated. The amount of energy to be committed in each market is expressed in terms of a fraction of the maximal energy output the generation portfolio would generate without failures in the period, i.e. 10MWh per hour of operation. The prices for the traded futures are also presented in Figure 9 except for the 2nd quarter future which is 45.63$/MWh and it is not shown as the optimal trade does not include this contract, presumably because the price is too low and it is better to wait for a better price in the spot market and sell in subsequent decisions. Likewise, the expected spot prices for each quarter are displayed, except for the last quarter which is 52.45$/MWh. Finally, the expected annual profits without considering fixed costs and the risk estimated by the RLS for the first rebalancing period are shown. Note that the expected profit is calculated considering that the following trading decisions are made taking into account the particular sample price realization, capturing the adaptation to the market developments. Thus, the rebalancing decisions for the second up to the last period are not unique.

5.3.1. Sensitivity to unit availability

In order to investigate the sensitivity of the optimal trading strategy to the unit availability, a second case was considered. Under these conditions, all the parameters are identical to the described base case, except for the failure and repair rates. Operation failure rate was set to 1/850h^{-1}, a reserve failure rate to 1/9850 h^{-1} and repair rate of 1/150h^{-1}. With these rates, the failure probability is 15% in operation and 1.5% in standby.

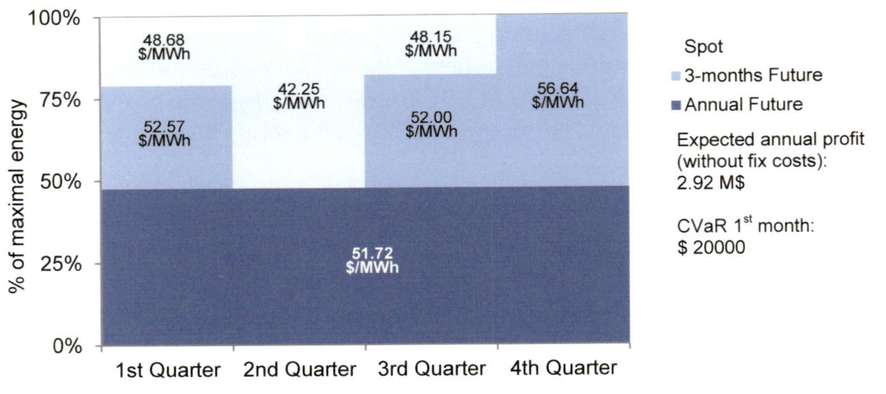

Figure 9. Optimal trading strategy for a 5x2MW generation portfolio

In Figure 10, the optimal trading strategy delivered by the ADP algorithm for the case of decreased unit availability is depicted. The differences on the sell strategy are evident for the annual future and the total amount of energy left to be sold in the spot market. Because of units have more frequent and longer random failures, a long-term commitment is avoided. However, the high prices for the fourth quarter push the sell in future markets up to 100%. In comparison to the base case results, the expected annual profit is slightly lower. The risk for the first month is irrelevant, because the forward commitment is around 50% and the probability of having more than two units (> 40%) unavailable is very low, leading to almost all unit failures can be covered by the remaining available units.

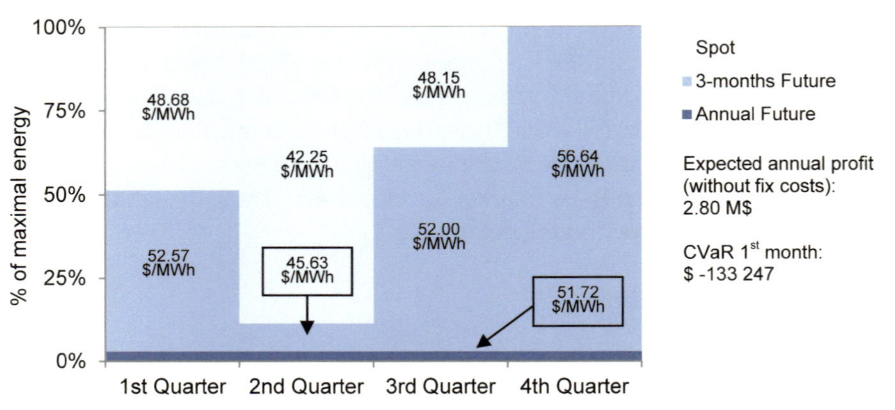

Figure 10. Optimal trading strategy for a 5x2MW generation portfolio with reduced availability

5.3.2. Sensitivity to transaction costs

The third case of study is identical to the base case but considering a transaction cost of 7% of the sold amount instead of 3%. The optimal trading policy for the case of increased transaction costs are shown in Figure 11. In this case, the generator sells only 1.32% of its capacity in an annual future contract due to the irreversibility introduced by the higher transactions costs and the larger contracting volume. Under these circumstances, the risk premium offered in the annual future price render insufficient for attracting the generator to enter in such a long-lasting commitment. On the other hand, the sell volume in quarterly futures is higher than in the base case, staying between 62% and 76% in a rather static trading policy. Under higher transaction fees, it is desirable to be able to rebalance the portfolio in future stages with smaller changes and hence smaller transaction costs. Expectedly, the expected profit is lower due to higher costs. The delivery risk for the first month is negligible, due to the fact that the most likely failures can still be covered by the remaining operative units without buying replacement power in the spot market.

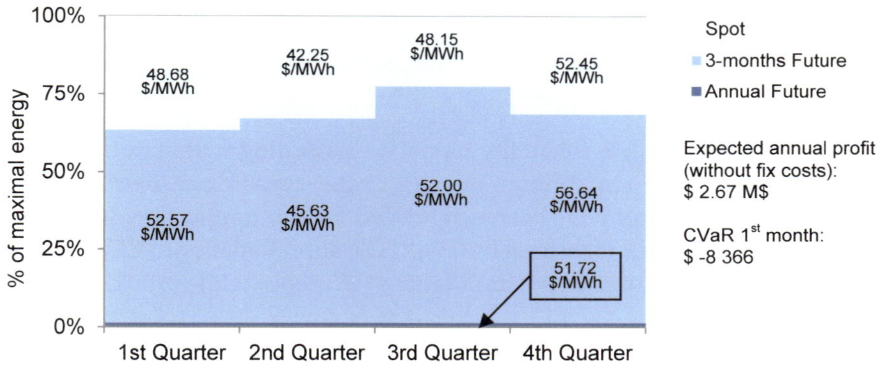

Figure 11. Optimal trading strategy for a 5x2MW generation portfolio with higher transaction costs

6. Conclusions

Optimal decision-making under uncertainty is a field of active research and uppermost relevance in science, engineering and computational finance. Conventional optimization approaches have difficulties and serious limitations for tackling high-dimensional problems often encountered in real world settings. Recent advances in operation research and computation technology opened new possibilities for approaching optimization problems that were considered intractable until recent times. This chapter presents an efficient Approximate Dynamic Programming algorithm for solving complex stochastic optimization problems and amenable for running in a distributed computing environment. The implemented ADP algorithm has been validated against conventional Dynamic Programming for a simple problem.

The proposed algorithm uses Monte Carlo simulation techniques combined with linear regression for successively approximating and refining the continuation and risk functions. A novel and efficient procedure for updating these functions, combining calculations of independent computing threads and without storing the entire datasets, is proposed. This feature enables exploiting the currently widespread multicore processor architectures and deploying the algorithm in large computation clusters.

In order to demonstrate the practicability of the envisioned approach, the proposed algorithm has been applied to find the optimal trading strategy of a power generation portfolio in forward and spot electricity markets. Power trading and risk management is currently a central activity of power companies running in liberalized electricity markets. The probability density functions of the profits a generator would make by participating in either the spot or the forward markets are extremely different. The forms and boundaries of these probability functions have drastic implications for risk when generators get involved in the spot or the forward markets. Generators can hedge price risk of spot markets by contracting forward, but by exposing themselves to delivery risk. Hence, the optimization problem is formulated as the maximization of the expected profit of the trading policy while the downside risk is constrained. For doing so, the generator selects and combines a portfolio of annual and quarterly forward contracts as well as involvement in the spot market. A frictional market with non-negligible transaction costs is considered.

A detailed chronological 4-state reliability model of generating units has been adopted for replicating stochastic behavior of random outages. Large stochastic ensembles of spot prices and forward prices time series have been synthetized for this application. In order to retain subadditivity, downside risk is measured by CVaR. The approximation of CVaR by a moment-based risk metric drastically improves computational efficiency while providing accurate and consistent risk estimations.

Applying ADP-based optimization techniques to electricity markets is a novel undertaking and opens a prospectively fertile avenue for research. In future works, further algorithmic enhancement are foreseen. Application of these methods for designing trading strategies that considers a larger set of available financial contracts as well as generation portfolios comprising renewable resources would provide results and findings of high practical significance.

Acknowledgements

This work was supported in part by the National Scientific and Technical Research Council (CONICET) and the Agency for Promotion of Science & Technology (ANPCyT), Argentina. The financial support of the German Academic Exchange Service (DAAD) is also gratefully acknowledged. The authors also thank the support of the colleagues at the Institute of Power Systems and Power Economics (IAEW), RWTH Aachen, Germany, especially Univ.-Prof. Dr.-Ing. Albert Moser as well as colleagues at Institute of Electrical Energy (IEE), National University of San Juan, Argentina.

http://dx.doi.org/10.5772/57466

Author details

Miguel Gil-Pugliese[1*] and Fernando Olsina[2]

*Address all correspondence to: miguel.gil.pugliese@gmail.com

1 Institute of Power Systems and Power Economics (IAEW), RWTH Aachen, Germany

2 Institute of Electrical Energy (IEE), National University of San Juan, Argentina

References

[1] Olsina F, Larisson C, Garcés F. Hedging and volumen risk in forward contracting [*in Spanish*]. Proceedings of XII ERIAC de CIGRE, Foz do Iguazu, Brazil, May 2007. paper C5.02.

[2] Powell WB. Approximate Dynamic Programming: Solving the curses of dimensionality. Wiley-Blackwell; 2007.

[3] Bertsekas DP,Tsitsiklis JN. Neuro-Dynamic Programming (Optimization and Neural Computation Series, 3). Athena Scientific; 1996.

[4] Bellman RE. Dynamic Programming. Princeton University Press, Princeton, NJ; 1957

[5] Stoft S. Power System Economics: Designing Markets for Electricity, New York, USA; IEEE Press/Wiley; 2002.

[6] Hadsell L, Maralhe A, Shawky, HA. Estimating the volatility of wholesale electricity spot prices in the US. The Energy Journal 2004; 25(4) 23-40.

[7] Bjorgan R, Liu C, Lawarrée J. Financial risk management in a competitive electricity market. IEEE Transactions on Power Systems 1999; 14(4) 1285-1291.

[8] Stoft S, Belden T, Goldman C, Pickle S. Primer on Electricity Futures and Other Derivatives. Lawrence Berkeley National Laboratory, University of California, LBNL-41098, 1998.

[9] Artzner P, Delbaen F, Eber JM, Heath D. Coherent measures of risk. Mathematical Finance 1999; 9(3) 203-228.

[10] Olsina F, Weber C. Stochastic simulation of spot power prices by spectral representation. IEEE Transactions on Power Systems 2009; 24(4) 1710-1719.

[11] Billinton R, Ge J. A comparison of four-state generating unit reliability models for peaking units. IEEE Transactions on Power Systems 2004; 19(2) 763–768.

[12] Billinton R, Allan R. Reliability evaluation of power systems. Plenum Press, New York, 1996.

[13] RockafellarR, Uryasev S, Zabarankin M. Generalized deviations in risk analysis. Finance and Stochastics 2006; 10(1) 51-74.

[14] Fischer T. Risk capital allocation by coherent risk measures based on one-sided moments. Insurance: Mathematics and Economics 2003; 32(1) 135-146.

[15] Pavlo AK. Higher moment coherent risk measures. Taylor & Francis, 2007.

[16] Rockafellar T, Uryasev S, ZabarankinM. Deviation measures in risk analysis and optimization. University of Florida, Department of Industrial & Systems Engineering Working Paper 2002.

Chapter 5

Using Dynamic Programming Based on Bayesian Inference in Selection Problems

Mohammad Saber Fallah Nezhad

Additional information is available at the end of the chapter

http://dx.doi.org/10.5772/57423

1. Introduction

An important subject in mathematical science that causes new improvements in data analysis is sequential analysis. In this type of analysis, the number of required observations is not fixed in advance, but is a variable and depends upon the values of the gathered observation. In sequential analysis, at any stage of data gathering process, to determine the number of required observations at the next stage, we analyze the data at hand and with respect to the obtained results, we determine how many more observations are necessary. In this way, the process of data gathering is cheaper and the information is used more effectively. In other words, the data gathering process in sequential analysis, in contrast to frequency analysis, is on-line. This idea caused some researches to conduct researches in various statistical aspects (Basseville and Nikiforov[1]).

In this chapter, using the concept of the sequential analysis approach, we develop an innovative Bayesian method designed specifically for the best solution in selection problem. The proposed method adopts the optimization concept of Bayesian inference and the uncertainty of the decision-making method in dynamic programming environment. The proposed algorithm is capable of taking into consideration the quality attributes of uncertain values in determining the optimal solution. Some authors have applied sequential analysis inference in combination with optimal stopping problem to maximize the probability of making correct decision. One of these researches is a new approach in probability distribution fitting of a given statistical data that Eshragh and Modarres [2] named it Decision on Belief (DOB). In this decision-making method, a sequential analysis approach is employed to find the best underlying probability distribution of the observed data. Moreover, Monfared and Ranaeifar [3] and Eshragh and Niaki [4] applied the DOB concept as a decision-making tool in some problems.

Since the idea behind the sequential analysis modeling is completely similar to the decision-making process of a human being in his life, it may perform better than available methods in decision-making problems. In these problems, when we want to make a decision, first we divide all of the probable solution space into smaller subspaces (the solution is one of the subspaces). Then based on our experiences, we assign a probability measure (belief) to each subspace, and finally we update the beliefs and make the decision.

2. An application to determine the best binomial distribution

In the best population selection problem, a similar decision-making process exits. First, the decision space can be divided into several subspaces (one for each population); second, the solution of the problem is one of the subspaces (the best population). Finally, we can assign a belief to each subspace where the belief denotes the performance of the population in term of its parameter. Based upon the updated beliefs in iterations of the data gathering process, we may decide which population possesses the best parameter value.

Consider n independent populations $P_1, P_2, \ldots, P_n$, where for each index $i = 1, 2, \ldots, n$, population P_i is characterized by the value of its parameter of interest p_i. Let $p_{[1]} \leq \ldots \leq p_{[n]}$ denote the ordered value of the parameters $p_1, \ldots, p_n$. If we assume that the exact pairing between the ordered and the unordered parameter is unknown, then, a population P_i with $p_i = p_{[n]}$ is called the best population.

There are many applications for the best population selection problem. As one application in supply chain environments, one needs to select the supplier among candidates that performs the best in terms of the quality of its products. As another example, in statistical analysis, we need to select a distribution among candidates that fits the collected observations the most. Selecting a production process that is in out-of-control state, selecting the stochastically optimum point of a multi-response problem, etc. are just a few of these applications.

The problem of selecting the best population was studied in papers by Bechhofer and Kulkarni [5] using the indifference zone approach and by Gupta and Panchapakesan [6] employing the best subset selection approach.

2.1. Belief and the approach of its improvement

Assume that there are n available *Binomial* populations and we intend to select the one with the highest probability of success. Furthermore, in each stage of the data gathering process and for each population, we take an independent sample of size m. Let us define $\alpha'_{i,t}$ and $\beta'_{i,t}$ to be the observed number of successes and failures of the ith *Binomial* population in the tth stage (sample) and $\alpha_{i,k}$ and $\beta_{i,k}$ to be the cumulative observed number of successes and failures of the ith *Binomial* population up to the kth stage (sample) respectively. In other words, $\alpha_{i,k} = \sum_{t=1}^{k} \alpha'_{i,t}$ and $\beta_{i,k} = \sum_{t=1}^{k} \beta'_{i,t}$. Then, in the kth stage defining $\overline{p_{i,k}}$ to be the estimated probability of success of the ith population obtained by $\frac{\alpha_{i,k}}{km}$, referring to Jeffrey's prior (Nair et al.[7]), for $\overline{p_{i,k}}$, we take a *Beta* prior distribution with parameters $\alpha_{i,0}$=0.5 and $\beta_{i,0}$=0.5. Then, using Bayesian inference, we can easily show that the posterior probability density function of $\overline{p_{i,k}}$ is

$$f(\overline{p_{i,k}}) = \frac{\Gamma(\alpha_{i,k} + \beta_{i,k} + 1)}{\Gamma(\alpha_{i,k} + 0.5)\Gamma(\beta_{i,k} + 0.5)} \overline{p_{i,k}}^{\alpha_{i,k} - 0.5} (1 - \overline{p_{i,k}})^{\beta_{i,k} - 0.5} \tag{1}$$

At stage k of the data gathering process, after taking a sample and observing the numbers of failures and successes, we update the probability distribution function of $\overline{p_{i,k}}$ for each population. To do this, define $B(\alpha_{i,k}, \beta_{i,k})$ as a probability measure (called belief) of the ith population to be the best one given $\alpha_{i,k}$ and $\beta_{i,k}$ as

$$B(\alpha_{i,k}, \beta_{i,k}) = \Pr\left\{ i^{\text{th}} \text{population is the best} \middle| \alpha_{i,k}, \beta_{i,k} \right\} \tag{2}$$

We then update the beliefs based on the values of $(\alpha_{i,k}, \beta_{i,k})$ for each population in iteration k. If we define $B(\alpha_{i,k-1}, \beta_{i,k-1})$ as the prior belief for each population, in order to update the posterior belief $B(\alpha_{i,k}, \beta_{i,k})$, since we may assume that the data are taken independently in each stage, we will have

$$\begin{aligned} B(\alpha_{i,k}, \beta_{i,k}) &= \\ &\frac{\Pr\left\{ i^{th} \text{ Population is the best} \middle| (\alpha_{i,k-1}, \beta_{i,k-1}) \right\} \Pr\left\{ (\alpha_{i,k}, \beta_{i,k}) \middle| i^{th} \text{ Population is the best} \right\}}{\sum_{j=1}^{n} \left[\Pr\left\{ j^{th} \text{ Population is the best} \middle| (\alpha_{j,k-1}, \beta_{j,k-1}) \right\} \Pr\left\{ (\alpha_{j,k}, \beta_{j,k}) \middle| j^{th} \text{ Population is the best} \right\} \right]} \\ &= \frac{B(\alpha_{i,k-1}, \beta_{i,k-1}) \Pr\left\{ (\alpha_{i,k}, \beta_{i,k}) \middle| i^{th} \text{ Population is the best} \right\}}{\sum_{j=1}^{n} \left[B(\alpha_{j,k-1}, \beta_{j,k-1}) \Pr\left\{ (\alpha_{j,k}, \beta_{j,k}) \middle| j^{th} \text{ Population is the best} \right\} \right]} \end{aligned} \tag{3}$$

From equation (3) we see that to update the beliefs, we need to evaluate $\Pr\{(\alpha_{i,k}, \beta_{i,k}) \mid i^{th} \text{ Population is the best}\}$; i=1, 2, ..., n in each decision-making stage. One way to do this is to use

$$\Pr\left\{ (\alpha_{i,k}, \beta_{i,k}) \middle| i^{th} \text{ Population is the best} \right\} = \frac{\overline{p_{i,k}}}{\sum_{j=1}^{n} \overline{p_{j,k}}} \tag{4}$$

Then, the probability given in equation (3) will increase when a better population is selected. In the next theorem, we will prove that when the number of decision-making stages goes to infinity this probability converges to one for the best population.

Theorem 1

If the ith population is the best, then $\underset{k\to\infty}{Lim} B(\alpha_{i,k}, \beta_{i,k}) = B_i = 1$.

In order to prove the theorem first we prove the following two lemmas.

Lemma 1:

Define a recursive sequence $\{R_{k,j}; j=1, 2, ..., l\}$ as

$$R_{k,j} = \begin{cases} \dfrac{c_j R_{k-1,j}}{\sum_{i=1}^{l} c_i R_{k-1,i}} & \text{for} \quad k = 1,2,3,... \\ P_j & \text{for} \quad k = 0 \end{cases} \tag{5}$$

where $c_1, c_2, ...,$ and c_l are different positive constants, $\sum_{j=1}^{l} P_j = 1$, and $P_j > 0$ Then, if $l_j = \underset{k\to\infty}{Lim}(R_{k,j})$, there exist at most one non-zero l_j.

Proof:

Suppose there are two nonzero $l_s > 0$ and $l_t > 0$. Taking the limit on $R_{k,j}$ as k goes to infinity we have

$$\underset{k\to\infty}{Lim}\ (R_{k,j}) = l_j = \underset{k\to\infty}{Lim}\left(\frac{c_j R_{k-1,j}}{\sum_{i=1}^{l} c_i R_{k-1,i}}\right) = \frac{c_j l_j}{\sum_{i=1}^{l} c_i l_i} \tag{6}$$

Now since $l_s > 0$ and $l_t > 0$, then by equation (6) we have

$$l_s = \frac{c_s l_s}{\sum_{i=1}^{l} c_i l_i} \Rightarrow c_s = \sum_{i=1}^{l} c_i l_i \text{ and } l_t = \frac{c_t l_t}{\sum_{i=1}^{l} c_i l_i} \Rightarrow c_t = \sum_{i=1}^{l} c_i l_i \tag{7}$$

In other words, we conclude $c_s = c_t$, which is a contradiction.

Lemma 2:

Sequence $R_{k,j}$ converges to one for $j=g$ and converges to zero for $j\neq g$, where g is an index for the maximum value of c_j.

Proof

From equation (6), we know that $\sum_{j=1}^{l} l_j=1$. Then by lemma 1, we have $l_i=1$ for only one i. Now suppose that $c_g=\max_{j\in\{1\ldots m\}}\{c_j\}$ and $g\neq i$. We will show that this is a contradiction. Consider $H_{k,i}=\frac{R_{k,g}}{R_{k,i}}$. By equation (5), we have $H_{k,i}=\frac{c_g}{c_i}H_{k-1,i}$. Since $H_{o,i}>0$ we will have

$$H_{k,i}=\frac{c_g}{c_i}H_{k-1,i}=\left(\frac{c_g}{c_i}\right)^k H_{0,i} \quad \Rightarrow \quad \underset{k\to\infty}{Lim}\left(H_{k,i}\right)=\infty \tag{8}$$

That is a contradiction because $\underset{k\to\infty}{Lim}(H_{k,i})=\frac{\underset{k\to\infty}{Lim}(R_{k,g})}{\underset{k\to\infty}{Lim}(R_{k,i})}=\frac{l_g}{l_i}=0$. So $l_g=1$

Now we are ready to prove the convergence property of the proposed method. Taking limit on both sides of equation (3), we will have

$$\underset{k\to\infty}{Lim}B\left(\alpha_{i,k},\beta_{i,k}\right)=B_i=\underset{k\to\infty}{Lim}\left[\frac{B\left(\alpha_{i,k-1},\beta_{i,k-1}\right)\Pr\left\{\left(\alpha_{i,k},\beta_{i,k}\right)\middle| i^{th}\text{ Population is the best}\right\}}{\sum_{j=1}^{n}\left[B\left(\alpha_{j,k-1},\beta_{j,k-1}\right)\Pr\left\{\left(\alpha_{j,k},\beta_{j,k}\right)\middle| j^{th}\text{ Population is the best}\right\}\right]}\right] \tag{9}$$

From the law of large numbers, we know that $\underset{k\to\infty}{Lim}\overline{p}_{j,k}=p_j$, where p_j is the probability of success of the jth population. Hence, using equation (7) we have $B_i=\frac{B_i p_i}{\sum_{j=1}^{n}B_j p_j}$. Then assuming population i is the best, i.e., it possesses the largest value of p_j's, by lemma 1 and 2 we conclude that $B_i=1$ and $\underset{j\neq i}{B_j}=0$. This concludes the convergence property of the proposed method.

In real-world applications, since there is a cost associated with the data gathering process we need to select the best population in a finite number of decision-making stages. In the next section, we present the proposed decision-making method in the form of a stochastic dynamic programming model in which there is a limited number of decision-making stages available to select the best population.

2.2. A dynamic programming approach

The proposed dynamic programming approach to model the decision-making problem of selecting the best Binomial population is similar to an optimal stopping problem.

Let us assume that to find the best population there is a limited number of stages (s) available. Then, the general framework of the decision-making process in each stage is proposed as:

1. Take an independent sample of size m from each population.
2. Calculate the posterior beliefs in terms of the prior beliefs using Bayesian approach.
3. Select the two biggest beliefs.
4. Based upon the values of the two biggest beliefs calculate the minimum acceptable belief.
5. If the maximum belief is more than the minimum acceptable belief, then we can conclude that the corresponding subspace is the optimal one. Otherwise, go to step 1.

In step 3 of the above framework, let populations i and j be the two candidates of being the best populations (it means that the beliefs of populations i and j are the two biggest beliefs) and we have s decision-making stages. If the biggest belief is more than a threshold (minimum acceptable belief)$d_{i,j}(s)$, $(0 \le d_{i,j}(s) \le 1)$, we select the corresponding subspace of that belief as the solution. Otherwise, the decision-making process continues by taking more observations. We determine the value of $d_{i,j}(s)$ such that the belief of making the correct decision is maximized. To do this suppose that for each population a new observation, $(\alpha_{j,k}, \beta_{j,k})$, is available at a given stage k. At this stage, we define $V(s, d_{i,j}(s))$ to be the expected belief of making the correct decision in s stages when two populations i and j are the candidates for the optimal population. In other words, if we let CS denote the event of making the correct decision, we define $V_{i,j}(s, d_{i,j}(s)) = E[B_{i,j}\{CS\}]$, where $B_{i,j}\{CS\}$ is the belief of making the correct decision. Furthermore, assume that the maximum of $V_{i,j}(s, d_{i,j}(s))$ occurs at $d_{i,j}^*(s)$. Then, we will have

$$V_{i,j}\left(s,d_{i,j}^{*}(s)\right) = \underset{d_{i,j}(s)}{Max}\left\{V_{i,j}\left(s,d_{i,j}(s)\right)\right\} = Max\left\{E\left[B_{i,j}\{CS\}\right]\right\} \tag{10}$$

We denote this optimal point by $V_{i,j}^*(s)$. In other words, $V_{i,j}^*(s) = V_{i,j}(s, d_{i,j}^*(s))$. Moreover, let us define S_i and S_j to be the state of selecting population i and j as the candidates for the optimal population, respectively, and $NS_{i,j}$ as the state of choosing neither of these population. Then, by conditioning on the above states, we have

$$\begin{aligned} &V_{i,j}^{*}(s) = Max\left\{E\left[B_{i,j}\{CS\}\right]\right\} = \\ &Max\left\{E\left[B_{i,j}\{CS|S_i\}B_{i,j}\{S_i\} + B_{i,j}\{CS|S_j\}B_{i,j}\{S_j\} + B_{i,j}\{CS|NS_{i,j}\}B_{i,j}\{NS_{i,j}\}\right]\right\} \end{aligned} \tag{11}$$

In order to evaluate$V_{i,j}^*(s)$, in what follows we will find the belief terms of equation (11).

a. $B_{i,j}\{CS \mid S_i\}$ and $B_{i,j}\{CS \mid S_j\}$

These are the beliefs of making the correct decision if population *i* or *j* is selected as the optimal population, respectively. To make the evaluation easier, we denote these beliefs by $B_{i,j}(i)$ and $B_{i,j}(j)$. Then, using equation (2) we have

$$B_{i,j}\{CS|S_i\} = B_{i,j}(i) = \frac{B(\alpha_{i,k-1},\beta_{i,k-1})\overline{p_{k,i}}}{B(\alpha_{i,k-1},\beta_{i,k-1})\overline{p_{k,i}} + B(\alpha_{j,k-1},\beta_{j,k-1})\overline{p_{k,j}}} \tag{12}$$

Similarly,

$$B_{i,j}\{CS|S_j\} = B_{i,j}(j) = \frac{B(\alpha_{j,k-1},\beta_{j,k-1})\overline{p_{k,j}}}{B(\alpha_{j,k-1},\beta_{j,k-1})\overline{p_{k,j}} + B(\alpha_{i,k-1},\beta_{i,k-1})\overline{p_{k,i}}} \tag{13}$$

b. $B_{i,j}\{S_i\}$ and $B_{i,j}\{S_j\}$

These are the beliefs of selecting population *i* or *j* as the optimal population, respectively. Regarding the decision-making strategy, we have:

$$B_{i,j}(i) = \max\left(B_{i,j}(i), B_{i,j}(j)\right) \text{ and } B_{i,j}(i) \geq d^*_{i,j}(s) \tag{14}$$

Hence, we define event S_i as

$$S_i \equiv \left\{B_{i,j}(i) = \max\left\{B_{i,j}(i), B_{i,j}(j)\right\}, B_{i,j}(i) \geq d^*_{i,j}(s)\right\} \tag{15}$$

Since $B_{i,j}(i) + B_{i,j}(j) = 1$ and that the beliefs are not negative we conclude $\max\{B_{i,j}(i),\ B_{i,j}(j)\} \geq 0.5$. Furthermore, since the decision making is performed based upon the maximum value of the beliefs, without interruption of assumptions, we can change the variation interval of $d^*_{i,j}(s)$ from [0,1] to [0.5,1]. Now by considering $d^*_{i,j}(s) \geq 0.5$ implicitly, we have $S_i \equiv \{B_{i,j}(i) \geq d^*_{i,j}(s)\}$. By similar reasoning $S_j \equiv \{B_{i,j}(j) \geq d^*_{i,j}(s)\}$. Hence

$$\begin{aligned} &B_{i,j}\{S_j\} = \Pr\left\{B_{i,j}(j) \geq d^*_{i,j}(s)\right\} = \\ &\Pr\left\{\frac{B(\alpha_{j,k-1},\beta_{j,k-1})\overline{p_{j,k}}}{B(\alpha_{i,k-1},\beta_{i,k-1})\overline{p_{i,k}} + B(\alpha_{j,k-1},\beta_{j,k-1})\overline{p_{j,k}}} \geq d^*_{i,j}(s)\right\} = \Pr\left\{\overline{p_{j,k}} \geq h\left(d^*_{i,j}(s)\right)\overline{p_{i,k}}\right\} = \Pr\left\{\frac{\overline{p_{j,k}}}{\overline{p_{i,k}}} \geq h\left(d^*_{i,j}(s)\right)\right\} \end{aligned} \tag{16}$$

In which, $h\left(d_{i,j}^{*}(s)\right)=\frac{d_{i,j}^{*}(s)B(\alpha_{i,k-1},\ \beta_{i,k-1})}{\left(1-d_{i,j}^{*}(s)\right)B(\alpha_{j,k-1},\ \beta_{j,k-1})}$.

To evaluate $\Pr\left\{\frac{\overline{p_{j,k}}}{\overline{p_{i,k}}}\geq h\left(d_{i,j}^{*}(s)\right)\right\}$ in equation (16), let $f_1(\overline{p_{j,k}})$ and $f_2(\overline{p_{i,k}})$ to be the probability distributions of $\overline{p_{j,k}}$ and $\overline{p_{i,k}}$, respectively. Then,

$$\begin{aligned}f_2(\overline{p_{i,k}})&=\frac{\Gamma(\alpha_{i,k}+\beta_{i,k}+1)}{\Gamma(\alpha_{i,k}+0.5)\Gamma(\beta_{i,k}+0.5)}\overline{p_{i,k}}^{\alpha_{i,k}-0.5}(1-\overline{p_{i,k}})^{\beta_{i,k}-0.5}\\ f_1(\overline{p_{j,k}})&=\frac{\Gamma(\alpha_{j,k}+\beta_{j,k}+1)}{\Gamma(\alpha_{j,k}+0.5)\Gamma(\beta_{j,k}+0.5)}\overline{p_{j,k}}^{\alpha_{j,k}-0.5}(1-\overline{p_{j,k}})^{\beta_{j,k}-0.5}\end{aligned} \tag{17}$$

Hence,

$$\begin{aligned}&\Pr\left\{\overline{p_{j,k}}\geq h\left(d_{i,j}^{*}(s)\right)\overline{p_{i,k}}\right\}=\int_0^1\int_{h\left(d_{i,j}^{*}(s)\right)\overline{p_{i,k}}}^{1}f_1\left(\overline{p_{j,k}}\right)f_2\left(\overline{p_{i,k}}\right)d\overline{p_{j,k}}d\overline{p_{i,k}}=\\ &\int_0^1\int_{h\left(d_{i,j}^{*}(s)\right)\overline{p_{i,k}}}^{1}A_i\overline{p_{i,k}}^{\alpha_{i,k}-0.5}(1-\overline{p_{i,k}})^{\beta_{i,k}-0.5}A_j\overline{p_{j,k}}^{\alpha_{j,k}-0.5}(1-\overline{p_{j,k}})^{\beta_{j,k}-0.5}d\overline{p_{j,k}}d\overline{p_{i,k}}\end{aligned} \tag{18}$$

where

$$A_i=\frac{\Gamma(\alpha_{i,k}+\beta_{i,k}+1)}{\Gamma(\alpha_{i,k}+0.5)\Gamma(\beta_{i,k}+0.5)},\ A_j=\frac{\Gamma(\alpha_{j,k}+\beta_{j,k}+1)}{\Gamma(\alpha_{j,k}+0.5)\Gamma(\beta_{j,k}+0.5)}. \tag{19}$$

By change of variables technique, we have:

$$\begin{aligned}&U=\frac{\overline{p_{j,k}}}{\overline{p_{i,k}}}\text{ and }V=\overline{p_{i,k}}\\ &f(U)=A_iA_jU^{\alpha_{i,k}-0.5}\int_0^1V^{\alpha_{i,k}+\alpha_{j,k}}(1-V)^{\beta_{i,k}-1}(1-UV)^{\beta_{j,k}-0.5}dV\\ &\Pr\left\{\frac{\overline{p_{j,k}}}{\overline{p_{i,k}}}\geq h\left(d_{i,j}^{*}(s)\right)\right\}=1-\int_0^{h\left(d_{i,j}^{*}(s)\right)}f(U)dU=1-F\left(h\left(d_{i,j}^{*}(s)\right)\right)\end{aligned} \tag{20}$$

For $B_{i,j}\{S_i\}$ we have

$$\begin{aligned}&B_{i,j}\{S_i\}=\Pr\left\{B_{i,j}(i)\geq d_{i,j}^{*}(s)\right\}=\\ &\Pr\left\{1-B_{i,j}(j)\geq d_{i,j}^{*}(s)\right\}=\Pr\left\{B_{i,j}(j)\leq 1-d_{i,j}^{*}(s)\right\}=F\left(h\left(1-d_{i,j}^{*}(s)\right)\right)\end{aligned} \tag{21}$$

http://dx.doi.org/10.5772/57423

c. $B_{i,j}\{CS \mid NS_{i,j}\}$

$B_{i,j}\{CS \mid NS_{i,j}\}$is the belief of making the correct decision when none of the subspaces i and j has been chosen as the optimal one. In other words, the maximum beliefs has been less than $d_{i,j}^{*}(s)$ and the process of decision-making continues to the next stage. In terms of stochastic dynamic programming approach, the belief of this event is equal to the maximum belief of making the correct decision in *(s-1)* stages. Since the value of this belief is discounted in the current stage, using discount factor α,

$$B_{i,j}\left\{CS \middle| NS_{i,j}\right\} = \alpha V_{i,j}^{*}(s-1) \tag{22}$$

Having all the belief terms of equation (11) evaluated in equations (12), (13), (14), (15), and (16), and knowing that by partitioning the state space we have $B_{i,j}\{NS_{i,j}\}=1-(B_{i,j}\{S_i\}+B_{i,j}\{S_j\})$, equation (11) can now be evaluated by substituting.

$$\begin{aligned} V_{i,j}^{*}(s) &= \max_{0.5\le d_{i,j}(s)\le 1}\{B_{i,j}(i)\Pr\{B_{i,j}(i)\ge d_{i,j}(s)\} + B_{i,j}(j)\Pr\{B_{i,j}(j)\ge d_{i,j}(s)\} \\ &+ B_{i,j}\{NS_{i,j}|CS\}\left(1-\Pr\{B_{i,j}(i)\ge d_{i,j}(s)\}-\Pr\{B_{i,j}(j)\ge d_{i,j}(s)\}\right)\} \\ &= \max_{0.5\le d_{i,j}(s)\le 1}\left\{B_{i,j}(i)\Pr\{B_{i,j}(i)\ge d_{i,j}(s)\} + B_{i,j}(j)\Pr\{B_{i,j}(j)\ge d_{i,j}(s)\} + \right. \\ &\left. \alpha V_{i,j}^{*}(s-1)\left(1-\Pr\{B_{i,j}(i)\ge d_{i,j}(s)\}-\Pr\{B_{i,j}(j)\ge d_{i,j}(s)\}\right)\right\} \end{aligned} \tag{23}$$

2.2.1. Making the decision

Assuming that for the two biggest beliefs we have $B_{i,j}(i)\ge B_{i,j}(j)$, equation (23) can be written as

$$\begin{aligned} V_{i,j}^{*}(s) &= \left(B_{i,j}(i)-\alpha V_{i,j}^{*}(s-1)\right)\Pr\{B_{i,j}(i)\ge d_{i,j}^{*}(s)\} + \\ &\left(B_{i,j}(j)-\alpha V_{i,j}^{*}(s-1)\right)\Pr\{B_{i,j}(j)\ge d_{i,j}^{*}(s)\} + \alpha V_{i,j}^{*}(s-1) \end{aligned} \tag{24}$$

For the decision-making problem at hand, three cases may happen

1. $B_{i,j}(i)<\alpha V_{i,j}^{*}(s-1)$:

In this case, both $\left(B_{i,j}(i)-\alpha V_{i,j}^{*}(s-1)\right)$ and $\left(B_{i,j}(j)-\alpha V_{i,j}^{*}(s-1)\right)$ are negative. Since we are maximizing $V_{i,j}(s, d_{i,j}(s))$, then the two probability terms in equation (24) must be minimized. This only happens when we let $d_{i,j}^{*}(s)=1$, making the probability terms equal to zero. Now since $B_{i,j}(i)<d_{i,j}^{*}(s)=1$, we continue to the next stage.

2. $B_{i,j}(j) > \alpha V^*_{i,j}(s-1)$:

In this case, $\left(B_{i,j}(i) - \alpha V^*_{i,j}(s-1)\right)$ and $\left(B_{i,j}(j) - \alpha V^*_{i,j}(s-1)\right)$ are both positive and to maximize $V_{i,j}(s, d_{i,j}(s))$ we need the two probability terms in equation (24) to be maximized. This only happens when we let $d^*_{i,j}(s) = 0.5$. Since $B_{i,j}(i) > d^*_{i,j}(s) = 0.5$, we select population i as the optimal subspace.

3. $B_{i,j}(j) \le \alpha V^*_{i,j}(s-1) \le B_{i,j}(i)$:

In this case, one of the probability terms in equation (24) has positive coefficient and the other has negative coefficient. In this case, in order to maximize $V_{i,j}(s, d_{i,j}(s))$ we take the derivative as follows.

Substituting equations (20) and (21) in equation (24) we have

$$\begin{aligned} V_{i,j}\left(s, d_{i,j}(s)\right) &= \left(B_{i,j}(i) - \alpha V^*_{i,j}(s-1)\right)\left\{F\left(h\left(1 - d_{i,j}(s)\right)\right)\right\} + \\ &\left(B_{i,j}(j) - \alpha V^*_{i,j}(s-1)\right)\left\{1 - F\left(h\left(d_{i,j}(s)\right)\right)\right\} + \alpha V^*_{i,j}(s-1) \end{aligned} \tag{25}$$

Thus following is obtained,

$$\begin{aligned} V_{i,j}{}^*(s) &= \left(B_{i,j}(i) - \alpha V^*_{i,j}(s-1)\right)\Pr\left\{\frac{\overline{p_{j,k}}}{\overline{p_{i,k}}} \le h\left(1 - d^*_{i,j}(s)\right)\right\} + \\ &\left(B_{i,j}(j) - \alpha V^*_{i,j}(s-1)\right)\Pr\left\{\frac{\overline{p_{j,k}}}{\overline{p_{i,k}}} \ge h\left(d^*_{i,j}(s)\right)\right\} + \alpha V^*_{i,j}(s-1) \end{aligned} \tag{26}$$

For determining $\Pr\left\{\frac{\overline{p_{j,k}}}{\overline{p_{i,k}}} \le h\left(1 - d^*_{i,j}(s)\right)\right\}$, first using an approximation, we assume that $\overline{p_{i,k}}$ is a constant number equal to its mean, then we have:

$$\begin{aligned} &\Pr\left\{\frac{\overline{p_{j,k}}}{\overline{p_{i,k}}} \le h\left(1 - d^*_{i,j}(s)\right)\right\} = \\ &\Pr\left\{\overline{p_{j,k}} \le \overline{p_{i,k}}h\left(1 - d^*_{i,j}(s)\right)\right\} = \int_0^{\overline{p_{i,k}}h\left(1 - d^*_{i,j}(s)\right)} f_1\left(p_{j,k}\right)dp_{j,k} = \\ &F_1\left(\overline{p_{i,k}}h\left(1 - d^*_{i,j}(s)\right)\right) \\ &\frac{\partial F_1\left(\overline{p_{i,k}}h\left(1 - d^*_{i,j}(s)\right)\right)}{\partial d^*_{i,j}(s)} = \frac{B\left(\alpha_{i,k-1}, \beta_{i,k-1}\right)}{\left(1 - d^*_{i,j}(s)\right)^2 B\left(\alpha_{j,k-1}, \beta_{j,k-1}\right)} f_1\left(\overline{p_{i,k}}h\left(1 - d^*_{i,j}(s)\right)\right) \end{aligned} \tag{27}$$

http://dx.doi.org/10.5772/57423

Also $\Pr\left\{\frac{\overline{p_{j,k}}}{\overline{p_{i,k}}} \geq h\left(d_{i,j}^{*}(s)\right)\right\}$ is obtained as follows,

$$\Pr\left\{\frac{\overline{p_{j,k}}}{\overline{p_{i,k}}} \geq h\left(d_{i,j}^{*}(s)\right)\right\} = \Pr\left\{\overline{p_{j,k}} \geq \overline{p_{i,k}} h\left(d_{i,j}^{*}(s)\right)\right\} = \int_{\overline{p_{i,k}} h\left(d_{i,j}^{*}(s)\right)}^{1} f_1\left(p_{j,k}\right) dp_{j,k} =$$
$$1 - F_1\left(\overline{p_{i,k}} h\left(d_{i,j}^{*}(s)\right)\right) \tag{28}$$
$$\frac{\partial\left(1 - F_1\left(\overline{p_{i,k}} h\left(d_{i,j}^{*}(s)\right)\right)\right)}{\partial d_{i,j}^{*}(s)} = \frac{-B\left(\alpha_{i,k-1}, \beta_{i,k-1}\right)}{\left(d_{i,j}^{*}(s)\right)^2 B\left(\alpha_{j,k-1}, \beta_{j,k-1}\right)} f_1\left(\overline{p_{i,k}} h\left(d_{i,j}^{*}(s)\right)\right)$$

Now following can be resulted,

$$\frac{\partial V_{i,j}\left(s, d_{i,j}(s)\right)}{\partial d_{i,j}(s)} = 0 \Rightarrow$$
$$\left(B_{i,j}(i) - \alpha V_{i,j}^{*}(s-1)\right) \frac{B\left(\alpha_{i,k-1}, \beta_{i,k-1}\right)}{\left(1 - d_{i,j}^{*}(s)\right)^2 B\left(\alpha_{j,k-1}, \beta_{j,k-1}\right)_1} f_1\left(\overline{p_{i,k}} h\left(1 - d_{i,j}^{*}(s)\right)\right) =$$
$$-\left(B_{i,j}(j) - \alpha V_{i,j}^{*}(s-1)\right) \frac{B\left(\alpha_{i,k-1}, \beta_{i,k-1}\right)}{\left(d_{i,j}^{*}(s)\right)^2 B\left(\alpha_{j,k-1}, \beta_{j,k-1}\right)} f_1\left(\overline{p_{i,k}} h\left(d_{i,j}^{*}(s)\right)\right) \Rightarrow \tag{29}$$
$$\frac{\left(B_{i,j}(i) - \alpha V_{i,j}^{*}(s-1)\right)}{-\left(B_{i,j}(j) - \alpha V_{i,j}^{*}(s-1)\right)} = \left(\frac{1 - d_{i,j}^{*}(s)}{d_{i,j}^{*}(s)}\right)^{2\left(\alpha_{i,k-1} + \beta_{i,k-1}\right)} \Rightarrow$$
$$d_{i,j}^{1}(s) = \frac{1}{\left(\frac{\left(B_{i,j}(i) - \alpha V_{i,j}^{*}(s-1)\right)}{-\left(B_{i,j}(j) - \alpha V_{i,j}^{*}(s-1)\right)}\right)^{\frac{1}{2\left(\alpha_{i,k-1} + \beta_{i,k-1}\right)}} + 1}$$

Now the approximate value of $d_{i,j}(s)$ say $d^{1}_{i,j}(s)$ is determined.

Second using another approximation, we assume that $\overline{p_{j,k}}$ is a constant number equal to its mean thus with similar reasoning, following is obtained:

$$d_{i,j}^{2}(s) = \frac{1}{\left(\frac{\left(B_{i,j}(j) - \alpha V_{i,j}^{*}(s-1)\right)}{-\left(B_{i,j}(i) - \alpha V_{i,j}^{*}(s-1)\right)}\right)^{\frac{1}{2\left(\alpha_{i,k-1} + \beta_{i,k-1}\right)}} + 1} \tag{30}$$

Therefore the approximate optimal value of $d_{i,j}^{*}(s)$ can be determined from following equation,

$$d_{i,j}^{*}(s) = Max\left\{d_{i,j}^{1}(s), d_{i,j}^{2}(s)\right\} \tag{31}$$

3. An application for fault detection and diagnosis in multivariate statistical quality control environments

3.1. Introduction

In this section, a heuristic threshold policy is applied in phase II of a control charting procedure to not only detect the states of a multivariate quality control system, but also to diagnose the quality characteristic(s) responsible for an out-of-control signal. It is assumed that the in-control mean vector and in-control covariance matrix of the process have been obtained in phase I.

3.2. Background

In a multivariate quality control environment, suppose there are m correlated quality characteristics whose means are being monitored simultaneously. Further, assume there is only one observation on the quality characteristics at each iteration of the data gathering process, where the goal is to detect the variable with the maximum mean shift. Let x_{ki} be the observation of the ith quality characteristic, $i=1, 2, ..., m$, at iteration k, $k=1, 2, ...$, and define the observation vector $\boldsymbol{x}_k = [x_{k1}, x_{k2}, ..., x_{km}]^T$ and observation matrix $\boldsymbol{O}_k = (\boldsymbol{x}_1, \boldsymbol{x}_2, ..., \boldsymbol{x}_k)$. After taking a new observation, $\boldsymbol{x}_k$, define $B_i(\boldsymbol{x}_k, \boldsymbol{O}_{k-1})$, the probability of variable i to be in an out-of-control state, as

$$B_i(\boldsymbol{x}_k, \boldsymbol{O}_{k-1}) = \Pr\{OOC_i | \boldsymbol{x}_k, \boldsymbol{O}_{k-1}\}, \tag{32}$$

where OOC stands for out-of-control. This probability has been called the belief of variable i to be in out-of-control condition given the observation matrix up to iteration $k-1$ and the observation vector obtained at iteration k.

Assuming the observations are taken independently at each iteration, to improve the belief of the process being in an out-of-control state at the kth iteration, based on the observation matrix $\boldsymbol{O}_{k-1}$ and the new observation vector $\boldsymbol{x}_k$, we have

$$\Pr\{\boldsymbol{x}_k | OOC_i, \boldsymbol{O}_{k-1}\} = \Pr\{\boldsymbol{x}_k | OOC_i\} \tag{33}$$

Then, using the Bayesian rule the posterior belief is:

$$B_i(\boldsymbol{x}_k,\boldsymbol{O}_{k-1}) = \Pr\{OOC_i|\boldsymbol{x}_k,\boldsymbol{O}_{k-1}\} = \frac{\Pr\{OOC_i,\boldsymbol{x}_k,\boldsymbol{O}_{k-1}\}}{\Pr\{\boldsymbol{x}_k,\boldsymbol{O}_{k-1}\}} = \frac{\Pr\{OOC_i,\boldsymbol{x}_k,\boldsymbol{O}_{k-1}\}}{\sum_{j=1}^{m}\Pr\{OOC_j,\boldsymbol{x}_k,\boldsymbol{O}_{k-1}\}} = \frac{\Pr\{OOC_i|\boldsymbol{O}_{k-1}\}\Pr\{\boldsymbol{x}_k|OOC_i,\boldsymbol{O}_{k-1}\}}{\sum_{j=1}^{m}\Pr\{OOC_j|\boldsymbol{O}_{k-1}\}\Pr\{\boldsymbol{x}_k|OOC_j,\boldsymbol{O}_{k-1}\}} \tag{34}$$

Since the goal is to detect the variable with the maximum mean shift, only one quality characteristic can be considered out-of-control at each iteration. In this way, there are $m-1$ remaining candidates for which $m-1$ quality characteristics are in-control. Hence, one can say that the candidates are mutually exclusive and collectively exhaustive. Therefore, using the Bayes' theorem, one can write equation (34) as

$$B_i(\boldsymbol{x}_k,\boldsymbol{O}_{k-1}) = \frac{\Pr\{OOC_i|\boldsymbol{O}_{k-1}\}\Pr\{\boldsymbol{x}_k|OOC_i\}}{\sum_{j=1}^{m}\Pr\{OOC_j|\boldsymbol{O}_{k-1}\}\Pr\{\boldsymbol{x}_k|OOC_j\}} = \frac{B_i(\boldsymbol{x}_{k-1},\boldsymbol{O}_{k-2})\Pr\{\boldsymbol{x}_k|OOC_i\}}{\sum_{j=1}^{m}B_j(\boldsymbol{x}_{k-1},\boldsymbol{O}_{k-2})\Pr\{\boldsymbol{x}_k|OOC_j\}} \tag{35}$$

When the system is in-control, we assume the m characteristics follow a multinormal distribution with mean vector $\boldsymbol{\mu} = [\mu_1, \mu_2, \ldots, \mu_m]^T$ and covariance matrix

$$\boldsymbol{\Sigma} = \begin{bmatrix} \sigma_1^2 & \sigma_{12} & \cdot & \sigma_{1m} \\ \sigma_{21} & \sigma_2^2 & \cdot & \sigma_{2m} \\ \cdot & \cdot & \cdot & \cdot \\ \sigma_{m1} & \sigma_{m2} & \cdot & \sigma_m^2 \end{bmatrix} \tag{36}$$

In out-of-control situations, only the mean vector changes and the probability distribution along with the covariance matrix remain unchanged. In latter case, equation (35) is used to calculate the probability of shifts in the process mean $\boldsymbol{\mu}$ at different iterations. Moreover, in order to update the beliefs at iteration k one needs to evaluate $\Pr\{\boldsymbol{x}_k \mid OOC_i\}$.

The term $\Pr\{\boldsymbol{x}_k \mid OOC_i\}$ is the probability of observing $\boldsymbol{x}_k$ if only the ith quality characteristic is out-of-control. The exact value of this probability can be determined using the multivariate normal density, $A\exp\left(-\frac{1}{2}(\boldsymbol{x}_k-\boldsymbol{\mu}_{1i})^T\boldsymbol{\Sigma}^{-1}(\boldsymbol{x}_k-\boldsymbol{\mu}_{1i})\right)$, where $\boldsymbol{\mu}_{1i}$ denotes the mean vector in which only the ith characteristic has shifted to an out-of-control condition and A is a known constant.

Since the exact value of the out-of-control mean vector $\boldsymbol{\mu}_{1i}$ is not known a priori, two approximations are used in this research to determine $\Pr\{\boldsymbol{x}_k \mid OOC_i\}$. Note that we do not want to determine the exact probability. Instead, the aim is to have an approximate probability (a belief) on each characteristic being out-of-control. In the first approximation method, define IC_i to be the event that all characteristics are in-control, and let $\Pr\{\boldsymbol{x}_k \mid IC_i\}$ be the conditional probability of observing $\boldsymbol{x}_k$ given all characteristics are in-control. Further, let $\boldsymbol{x}_k^{'}=[\mu_{01}, \ldots, x_{ki}, \mu_{0i+1}, \ldots, \mu_{0m}]^T$ in the aforementioned multivariate normal density, so that $\Pr\{\boldsymbol{x}_k \mid IC_i\}$ can be approximately evaluated using $\Pr\{\boldsymbol{x}_k \mid IC_i\}=\Pr\{\boldsymbol{x}'_k \mid IC_i\}$, where $\Pr\left\{\boldsymbol{x}'_k \mid IC_i\right\}=A\,\exp\left(-{}^1\!/_2(\boldsymbol{x}_k^{'}-\boldsymbol{\mu}_0)^T\boldsymbol{\Sigma}^{-1}(\boldsymbol{x}_k^{'}-\boldsymbol{\mu}_0)\right)$. Note that this evaluation is proportional to $\exp\left(-{}^1\!/_2\left(\frac{x_{ki}-\mu_{0i}}{\sigma_i}\right)^2\right)$, and since it is assumed that characteristic i is under control, no matter the condition of the other characteristics, this approximation is justifiable.

In the second approximation method, we assume $\Pr\left\{\boldsymbol{x}_k \mid OOC_i\right\} \propto \frac{1}{\Pr\{\boldsymbol{x}_k \mid IC_i\}}$. Although it is obvious that $\Pr\{\boldsymbol{x}_k \mid OOC_i\}$ is not equal to $\frac{1}{\Pr\{\boldsymbol{x}_k \mid IC_i\}}$, since we only need a belief function to evaluate $\Pr\{\boldsymbol{x}_k \mid OOC_i\}$ and also we do not know the exact value of out-of-control mean vector, this approximation is just used to determine$\Pr\{\boldsymbol{x}_k \mid OOC_i\}$. Moreover, it can be easily seen that the closer the value of the ith characteristic is to its in-control mean the smaller is $\Pr\{\boldsymbol{x}_k \mid OOC_i\}$ as expected. We thus let

$$\Pr\{\boldsymbol{x}_k|OOC_i\} \propto \frac{1}{\Pr\{\boldsymbol{x}_k|IC_i\}} = R\exp\left({}^1\!/_2\left(\frac{x_{ki}-\mu_{0i}}{\sigma_i}\right)^2\right) \quad ; \quad i=1,2,\ldots,m, \tag{37}$$

where R is a sufficiently big constant number to ensure the above definition is less than one. The approximation to $\Pr\{\boldsymbol{x}_k \mid OOC_i\}$ in equation (37) has the following two properties:

- It does not require the value of out-of-control means to be known.
- The determination of a threshold for the decision-making process (derived later) will be easier.

Niaki and Fallahnezhad [8] defined another equation for the above conditional probability and showed that if a shift occurs in the mean of variable i, then$\lim_{k\to\infty} B_i(\boldsymbol{x}_k, \boldsymbol{O}_{k-1})=B_i=1$. They proposed a novel method of detection and classification and used simulation to compare its performances with that of existing methods in terms of the average run length for different mean shifts. The results of the simulation study were in favor of their proposed method in almost all shift scenarios. Besides using a different equation, the main difference between the current research and Niaki and Fallahnezhad [8] is that the current work develops a novel heuristic threshold policy, in which to save sampling cost and time or when these factors are constrained, the number of the data gathering stages is limited.

3.3. The proposed procedure

Assuming a limited number of the data gathering stages, N, to detect and diagnose characteristic(s), a heuristic threshold policy-based model is developed in this Section. The framework of the proposed decision-making process follows.

Step I

Define $i=1, 2, \ldots, m$ as the set of indices for the characteristics, all of which having the potential of being out-of-control.

Step II

Using the maximum entropy principle, initialize $B_i(O_0)=1/m$ as the prior belief of the ith variable to be out-of-control. In other words, at the start of the decision-making process all variables have an equal chance of being out-of-control. Set the discount rateα, the maximum probability of correct selection when N decision making stages remains$V(N)$, and the maximum number of decision making stages$N$.

Step III

Set $k=0$

Step IV

Obtain an observation of the process.

Step V

Estimate the posterior beliefs, $B_i(O_k)$ (for$i=1, 2, \ldots, m$), using equation (35).

Step VI

Obtain the order statistics on the posterior beliefs $B_i(O_k)$ such that

$B_{(1)}(O_k)<B_{(2)}(O_k)<\ldots<B_{(m)}(O_k)$.

Furthermore, let $B_{gr}(O_k)=B_{(m)}(O_k)$ and $B_{sm}(O_k)=B_{(m-1)}(O_k)$.

Step VII

Assume the variables with the indices $i=gr$ and $j=sm$ are the candidates of being out-of-control, where N decision-making steps are available. Define $V(N, d_{i,j}(k))$ the probability of correct choice between the variables i and j, where $d_{i,j}(k)$ is the acceptable belief. Also, define CS the event of correct selection and event $E_{i,j}$ the existence of two out-of-control candidate variables i and j. Then, we have:

$$V\left(N, d_{i,j}(k)\right)=\Pr\{CS | E_{i,j}\} \triangleq \Pr_{i,j}\{CS\} \tag{38}$$

where " $\triangleq$ " means "defined as."

Assuming $d^*_{i,j}(k)$ the maximum point of $V(N, d_{i,j}(k))$, called the minimum acceptable belief, we have

$$V\left(N,d^*_{i,j}(k)\right) \triangleq V^*_{i,j}(N) = \underset{d_{i,j}(k)}{Max}\left\{V\left(N,d_{i,j}(k)\right)\right\} \triangleq Max\left\{\Pr_{i,j}\left\{CS\right\}\right\} \tag{39}$$

Let S_i and S_j be the event of selecting i and j as the out-of-control variables, respectively, and $NS_{i,j}$ be the event of not selecting any. Then, by conditioning on the probability, we have:

$$\begin{aligned} &V^*_{i,j}(N) = Max\left\{\Pr_{i,j}\left\{CS\right\}\right\} = \\ &Max\left\{\Pr_{i,j}\left\{CS|S_i\right\}\Pr_{i,j}\left\{S_i\right\} + \Pr_{i,j}\left\{CS|S_j\right\}\Pr_{i,j}\left\{S_j\right\} + \Pr_{i,j}\left\{CS|NS_{i,j}\right\}\Pr_{i,j}\left\{NS_{i,j}\right\}\right\} \end{aligned} \tag{40}$$

At the kth iteration, the conditional bi-variate distribution of the sample means for variables *gr* and *sm*, i.e, $X_{k,\, j=gr,sm} \mid X_{k,\, j\neq gr,sm}$, is determined using the conditional property of multivariate normal distribution given in appendix 1. Moreover, knowing $E(x_{k,j})=\mu_j$ and evaluating the conditional mean and standard deviation (see appendix 1) results in

$$E\left(X_{k,i}|X_{k,j}\right) = \mu_i + \rho\frac{\sigma_i}{\sigma_j}\left(X_{k,j} - \mu_j\right) \tag{41}$$

and

$$E\left(X_{k,i}|X_{k,j}\right) = \mu_i + \rho\frac{\sigma_i}{\sigma_j}\left(X_{k,j} - \mu_j\right) \tag{42}$$

Based on the decomposition method of Mason et al. [9], define statistics $T_{k,j}$ and $T_{k,i\mid j}$ as

$$T_{k,j} = \left(\frac{X_{k,j} - \mu_j}{\sigma_j}\right) \tag{43}$$

$$T_{k,i|j} = \left(\frac{X_{k,i} - E\left(X_{k,i}|X_{k,j}\right)}{\sigma_{X_{k,i}|X_{k,j}}}\right) \tag{44}$$

Thus, when the process is in-control, the statistics $T_{k,j}$ and $T_{k,i\mid j}$ follow a standard normal distribution [9].

Now, let $B_{i,j}(i; x_k, O_{k-1})$ denote the probability of correct selection conditioned on selecting i as the out-of-control variable. Hence,

$$\begin{aligned} B_{i,j}(i;x_k,O_{k-1}) &= \frac{B_i(O_{k-1})e^{(0.5T_{k,i|j})^2}}{B_j(O_{k-1})e^{(0.5T_{k,j})^2} + B_i(O_{k-1})e^{(0.5T_{k,i|j})^2}} \\ B_{i,j}(j;x_k,O_{k-1}) &= \frac{B_j(O_{k-1})e^{(0.5T_{k,j})^2}}{B_j(O_{k-1})e^{(0.5T_{k,j})^2} + B_i(O_{k-1})e^{(0.5T_{k,i|j})^2}} \end{aligned} \tag{45}$$

Then, the probability measure $\Pr_{i,j}\{CS \mid S_i\}$ is calculated using the following equation,

$$\Pr_{i,j}\{CS|S_i\} = B_{i,j}(i;x_k,O_{k-1}) \tag{46}$$

The probability measure $\Pr_{i,j}\{S_i\}$ is defined as the probability of selecting variable i to be out-of-control. Regarding to the explained strategy, we have:

$$S_i \equiv \left\{ B_{i,j}(i;x_k,O_{k-1}) = Max\begin{Bmatrix} B_{i,j}(i;x_k,O_{k-1}), \\ B_{i,j}(j;x_k,O_{k-1}) \end{Bmatrix}, B_{i,j}(i;x_k,O_{k-1}) \geq d_{i,j}(k) \right\} \tag{47}$$

Since $B_{i,j}(i; x_k, O_{k-1}) + B_{i,j}(j; x_k, O_{k-1}) = 1$ and the value of beliefs are not negative, we conclude

$$\max\left\{B_{i,j}(i;x_k,O_{k-1}), B_{i,j}(j;x_k,O_{k-1})\right\} \geq 0.5 \tag{48}$$

Without interruption of assumptions, we can change the variation interval of $d_{i,j}(k)$ from [0,1] to [0.5,1]. Hence,

$$S_i \equiv \left\{B_{i,j}(i;x_k,O_{k-1}) \geq d_{i,j}(k)\right\} \tag{49}$$

By similar reasoning, we have:

$$S_j \equiv \left\{B_{i,j}(j;x_k,O_{k-1}) \geq d_{i,j}(k)\right\} \tag{50}$$

The term $\Pr_{i,j}\{CS \mid NS_{i,j}\}$ denotes the probability of correct selection conditioned on excluding the candidates i and j as the solution. In other words, the maximum belief has been less than the threshold (minimum acceptable belief) $d^*_{i,j}(k)$ and the decision making process continues to the next stage. In terms of stochastic dynamic programming approach, the probability of this event is equal to the maximum probability of correct selection when there are $N-1$ stages remaining. The discounted value of this probability in the current stage using the discount factor α equals to $\alpha\, V_{i,j}(N-1)$. Further, since we partitioned the decision space into events $\{NS_{i,j}; S_i; S_j\}$, we have:

$$\Pr_{i,j}\left\{NS_{i,j}\right\} = 1 - \left(\Pr_{i,j}\left\{S_i\right\} + \Pr_{i,j}\left\{S_j\right\}\right) \tag{51}$$

Now we evaluate $V^*_{i,j}(N)$ as follows,

$$\begin{aligned} V^*_{i,j}(N) &= \\ &\underset{0.5\le d_{i,j}(k)\le 1}{Max}\left\{\begin{array}{l} B_{i,j}\left(i;x_k,O_{k-1}\right)\Pr_{i,j}\left\{B_{i,j}\left(i;x_k,O_{k-1}\right)\ge d_{i,j}(k)\right\}+ \\ B_{i,j}\left(j;x_k,O_{k-1}\right)\Pr_{i,j}\left\{B_{i,j}\left(j;x_k,O_{k-1}\right)\ge d_{i,j}(k)\right\}+ \\ \Pr_{i,j}\left\{NS_{i,j}\middle|CS\right\}\left(\begin{array}{l} 1-\Pr_{i,j}\left\{B_{i,j}(i;x_k,O_{k-1})\ge d_{i,j}(k)\right\}- \\ \Pr\left\{B_{i,j}(j;x_k,O_{k-1})\ge d_{i,j}(k)\right\} \end{array}\right) \end{array}\right\} \\ &= \underset{0.5\le d_{i,j}(k)\le 1}{Max}\left\{\begin{array}{l} B_{i,j}\left(i;x_k,O_{k-1}\right)\Pr_{i,j}\left\{B_{i,j}\left(i;x_k,O_{k-1}\right)\ge d_{i,j}(k)\right\}+ \\ B_{i,j}\left(j;x_k,O_{k-1}\right)\Pr_{i,j}\left\{B_{i,j}\left(j;x_k,O_{k-1}\right)\ge d_{i,j}(k)\right\}+ \\ \alpha V^*_{i,j}\left(N-1\right)\left(\begin{array}{l} 1-\Pr_{i,j}\left\{B_{i,j}\left(i;x_k,O_{k-1}\right)\ge d_{i,j}(k)\right\}- \\ \Pr\left\{B_{i,j}\left(j;x_k,O_{k-1}\right)\ge d_{i,j}(k)\right\} \end{array}\right) \end{array}\right\} \end{aligned} \tag{52}$$

In other words,

$$V^*_{i,j}(N) = \underset{0.5\le d_{i,j}(k)\le 1}{Max}\left\{\begin{array}{l} B_{i,j}\left(i;O_k\right)\Pr_{i,j}\left\{B_{i,j}\left(i;O_k\right)\ge d_{i,j}(k)\right\}+ \\ B_{i,j}\left(j;O_k\right)\Pr_{i,j}\left\{B_{i,j}\left(j;O_k\right)\ge d_{i,j}(k)\right\}+ \\ \alpha V^*_{i,j}(N-1)\left(\begin{array}{l} 1-\Pr_{i,j}\left\{B_{i,j}\left(i;O_k\right)\ge d_{i,j}(k)\right\} \\ -\Pr\left\{B_{i,j}\left(j;O_k\right)\ge d_{i,j}(k)\right\} \end{array}\right) \end{array}\right\} \tag{53}$$

The method of evaluating the minimum acceptable belief $d^*_{gr,sm}(k)$ is given in Appendix 2.

Step VIII: The Decision Step

If the belief $B_{gr,sm}(gr; x_k, O_{k-1})$ in the candidate set (sm, gr) is equal to or greater than $d^*_{gr,sm}(k)$ then choose the variable with index gr to be out-of-control. In this case, the decision-making process ends. Otherwise, without having any selection at this stage, obtain another observation, lower the number of remaining decision-stages to $N-1$, set $k=k+1$, and return to step **V** above. The process will continue until either the stopping condition is reached or the number of stages is finished. The optimal strategy with N decision-making stages that maximizes the probability of correct selection would be resulted from this process.

In what follows, the procedure to evaluate $V^*_{i,j}(N)$ of equation (53) is given in detail.

3.4. Method of evaluating $V^*_{i,j}(N)$

Using $d^*_{i,j}(k)$ as the minimum acceptable belief, from equation (53) we have

$$\begin{aligned} V^*_{i,j}(N) = &\left(B_{i,j}(i;O_k) - \alpha V^*_{i,j}(N-1)\right)\Pr\left\{B_{i,j}(i;O_k) \ge d^*_{i,j}(k)\right\} + \\ &\left(B_{i,j}(j;O_k) - \alpha V^*_{i,j}(N-1)\right)\Pr\left\{B_{i,j}(j;O_k) \ge d^*_{i,j}(k)\right\} + \alpha V^*_{i,j}(N-1) \end{aligned} \tag{54}$$

Then, for the decision-making problem at hand, three cases may occur

1. $B_{i,j}(i;O_k) < \alpha V^*_{i,j}(N-1)$

In this case, both $\left(B_{i,j}(i;O_k) - \alpha V^*_{i,j}(N-1)\right)$ and $\left(B_{i,j}(i;O_k) - \alpha V^*_{i,j}(N-1)\right)$ are negative. Since we are maximizing $V_{i,j}(N, d_{i,j}(k))$, the two probability terms in equation (54) must be minimized. This only happens when $d^*_{i,j}(k)=1$, making the probability terms equal to zero. In other words, since $B_{i,j}(i;O_k) < d^*_{i,j}(k)=1$, we continue to the next stage.

2. $B_{i,j}(j;O_k) > \alpha V^*_{i,j}(s-1)$

In this case, $\left(B_{i,j}(i;O_k) - \alpha V^*_{i,j}(N-1)\right)$ and $\left(B_{i,j}(i;O_k) - \alpha V^*_{i,j}(N-1)\right)$ are both positive and to maximize $V_{i,j}(N, d_{i,j}(k))$ we need the two probability terms in equation (54) to be maximized. This only happens when $d^*_{i,j}(k)=0.5$. In other words, since $B_{i,j}(i;O_k) > d^*_{i,j}(k)=0.5$, the variable with the index i is selected.

3. $B_{i,j}(j;O_k) < \alpha V^*_{i,j}(N-1) < B_{i,j}(i;O_k)$

In this case, one of the probability terms in equation (54) has a positive and the other a negative coefficient. Then, in order to maximize $V_{i,j}(N, d_{i,j}(k))$, the first derivative on $d_{i,j}(k)$ must be equated to zero. To do this, define $h\left(d_{gr,sm}(k)\right)$ and $r\left(d_{gr,sm}(k)\right)$ as follows:

$$h\left(d_{gr,sm}(k)\right)=\frac{d_{gr,sm}(k)B_{gr}\left(gr,O_{k-1}\right)}{\left(1-d_{gr,sm}(k)\right)B_{sm}\left(sm,O_{k-1}\right)} \tag{55}$$

$$r\left(d_{gr,sm}(k)\right)=\frac{d_{gr,sm}(k)B_{sm}\left(sm,O_{k-1}\right)}{\left(1-d_{gr,sm}(k)\right)B_{gr}\left(gr,O_{k-1}\right)} \tag{56}$$

We first present the method of evaluating $\Pr\{B_{gr,sm}(sm;O_k)\geq d_{gr,sm}(k)\}$ as follows.

$$\begin{aligned}&\Pr\left\{B_{gr,sm}\left(sm;O_k\right)\geq d_{gr,sm}(k)\right\}=\\&\Pr\left\{\frac{B_{sm}\left(sm,O_{k-1}\right)e^{(T_{k,gr|sm})^2}}{B_{sm}\left(sm,O_{k-1}\right)e^{(T_{k,gr|sm})^2}+B_{gr}\left(gr,O_{k-1}\right)e^{(T_{k,sm})^2}}\geq d_{gr,sm}(k)\right\}\\&=\Pr\left\{e^{(T_{k,gr|sm})^2}\geq h\left(d_{gr,sm}(k)\right)e^{(T_{k,sm})^2}\right\}\end{aligned} \tag{57}$$

Then, the method of evaluating probability terms in equation (57) is given in appendix 2.

With similar reasoning, we have,

$$\Pr\left\{B_{gr,sm}\left(gr;O_k\right)\geq d_{gr,sm}(k)\right\}=\Pr\left\{e^{(T_{k,gr|sm})^2}\geq r\left(d_{gr,sm}(k)\right)e^{(T_{k,sm})^2}\right\} \tag{58}$$

The method of determining the minimum acceptable belief is given in appendix 2.

4. An application for fault detection in uni-variate statistical quality control environments

In a uni-variate quality control environment, if we limit ourselves to apply a control charting method, most of the information obtained from data behavior will be ignored. The main aim of a control charting method is to detect quickly undesired faults in the process. However, we may calculate the belief for the process being out-of-control applying Bayesian rule at any iteration in which some observations on the quality characteristic are gathered. Regarding these beliefs and a stopping rule, we may find and specify a control threshold for these beliefs and when the updated belief in any iteration is more than this threshold, an out-of-control signal is observed.

In Decision on Beliefs, first, all probable solution spaces will be divided into several candidates (the solution is one of the candidates), then a belief will be assigned to each candidate considering our experiences and finally, the beliefs are updated and the optimal decision is selected based on the current situation. In a SPC problem, a similar decision-making process exits. First, the decision space can be divided into two candidates; an in-control or out-of-control production process. Second, the problem solution is one of the candidates (in-control or out-of-control process). Finally, a belief is assigned to each candidate so that the belief shows the probability of being a fault in the process. Based upon the updated belief, we may decide about states of the process (in-control or out-of-control process).

4.1. Learning — The beliefs and approach for its improvement

For simplicity, individual observation on the quality characteristic of interest in any iteration of data gathering process was gathered. At iteration k of data gathering process, $O_k = (x_1, x_2, \ldots\ldots, x_k)$was defined as the observation vector where resemble observations for previous iterations 1, 2, …, k. After taking a new observation, O_{k-1} the belief of being in an out-of-control state is defined as $B(x_k, O_{k-1}) = \Pr\{Out - of - control \mid x_k, O_{k-1}\}$. At this iteration, we want to update the belief of being in out-of-control state based on observation vector O_{k-1} and new observation x_k. If we define $B(O_{k-1}) = B(x_{k-1}, O_{k-2})$ as the prior belief of an out-of-control state, in order to update the posterior belief $B(x_k, O_{k-1})$, since we may assume that the observations are taken independently in any iteration, then we will have

$$\Pr\left\{x_k \middle| Out - of - control, O_{k-1}\right\} = \Pr\left\{x_k \middle| Out - of - control\right\} \tag{59}$$

With this feature, the updated belief is obtained using Bayesian rule:

$$\begin{aligned} B(x_k, O_{k-1}) &= \Pr\{Out - of - control | x_k, O_{k-1}\} = \frac{\Pr\{Out - of - control, x_k | O_{k-1}\}}{\Pr\{x_k | O_{k-1}\}} \\ &= \frac{\Pr\{Out - of - control | O_{k-1}\} \Pr\{x_k | Out - of - control, O_{k-1}\}}{\Pr\{x_k | O_{k-1}\}} \end{aligned} \tag{60}$$

Since in-control or out-of-control state partition the decision space, we can write equation (60) as

$$\begin{aligned} B(x_k, O_{k-1}) &= \frac{\Pr\{Out - of - control | O_{k-1}\} \Pr\{x_k | Out - of - control\}}{\Pr\{Out - of - control | O_{k-1}\} \Pr\{x_k | Out - of - control\} + \Pr\{In - control | O_{k-1}\} \Pr\{x_k | In - control\}} \\ &= \frac{B(O_{k-1}) \Pr\{x_k | Out - of - control\}}{B(O_{k-1}) \Pr\{x_k | Out - of - control\} + (1 - B(O_{k-1})) \Pr\{x_k | In - control\}} \end{aligned} \tag{61}$$

Assuming the quality characteristic of interest follows a normal distribution with mean μ and variance σ^2, we use equation (61) to calculate both beliefs for occurring positive or negative shifts in the process mean μ.

- Positive shifts in the process mean

The values of $B^{+}(O_k)$, showing the probability of occurring a positive shift in the process mean, will be calculated applying equation (61) recursively. $\Pr\{x_k \mid In-control\}$ is defined by the following equation,

$$\Pr\left\{x_k \middle| In-control\right\} = 0.5 \tag{62}$$

For positive shift, the probability of being a positive shift in the process at iteration k, $\Pr\{x_k \mid Out-of-control\}$, is calculated using equation (63).

$$\Pr\{x_k | Out-of-control\} = \varphi(x_k) \tag{63}$$

where $\varphi(x_k)$ is the cumulative probability distribution function for the normal distribution with mean μ and variance σ^2. Above probabilities are not exact probabilities and they are a kind of belief function to ascertain good properties for $B^{+}(O_k)$

Therefore $B^{+}(O_k)$ is determined by the following equation,

$$B^{+}(O_k) = \frac{B^{+}(O_{k-1})\varphi(x_k)}{B^{+}(O_{k-1})\varphi(x_k) + 0.5\left(1 - B^{+}(O_{k-1})\right)} \tag{64}$$

- Negative shifts in the process mean

The values of $B^{-}(O_k)$ denotes the probability of being a negative shift in the process mean that is calculated using equation (61) recursively. In this case, $\Pr\{x_k \mid In-control\}$ is defined by the following equation,

$$\Pr\left\{x_k \middle| In-control\right\} = 0.5 \tag{65}$$

Also is $\Pr\{x_k \mid Out-of-control\}$ calculated using equation (66).

$$\Pr\{x_k | Out-of-control\} = 1 - \varphi(x_k) \tag{66}$$

Thus $B^{-}(O_k)$ is determined by the following equation,

$$B^{-}(O_k) = \frac{B^{-}(O_{k-1})\left(1 - \varphi(x_k)\right)}{B^{-}(O_{k-1})\left(1 - \varphi(x_k)\right) + 0.5\left(1 - B^{-}(O_{k-1})\right)} \tag{67}$$

4.2. A decision on beliefs approach

We present a decision making approach in terms of Stochastic Dynamic Programming approach. Presented approach is like an optimal stopping problem.

Suppose n stages for decision making is remained and two decisions are available.

- A positive shift is occurred in the process mean
- No positive shift is occurred in the process mean

Decision making framework is as follows:

- Gather a new observation.
- Calculate the posterior Beliefs in terms of prior Beliefs.
- Order the current Beliefs as an ascending form and choose the maximum.
- Determine the value of the minimum acceptable belief ($d^{+}(n)$ is the minimum acceptable belief for detecting the positive shift and $d^{-}(n)$ is the least acceptable belief for detecting the negative shift)
- If the maximum Belief was more than the minimum acceptable belief, $d^{+}(n)$, select the belief candidate with maximum value as a solution else go to step 1.
- In terms of above algorithm, the belief with maximum value is chosen and if this belief was more than a control threshold like $d^{+}(n)$, the candidate of that Belief will be selected as optimal candidate else the sampling process is continued. The objective of this model is to determine the optimal values of $d^{+}(n)$. The result of this process is the optimal strategy with n decision making stages that maximize the probability of correct selection.

Suppose new observation x_k is gathered. (k is the number of gathered observations so far). $V(n,\ d^{+}(n))$ is defined as the probability of correct selection when n decision making stages are remained and we follow $d^{+}(n)$ strategy explained above also $V(n)$ denotes the maximum value of $V(n,\ d^{+}(n))$ thus,

$$V(n) = \underset{d^{+}(n)}{Max}\left\{V\left(n, d^{+}(n)\right)\right\} \tag{68}$$

CS is defined as the event of correct selection. S_1 is defined as selecting the out-of-control condition (positive shift) as an optimal solution and S_2 is defined as selecting the in-control condition as an optimal decision and NS is defined as not selecting any candidate in this stage.

Hence, using the total probability law, it is concluded that:

$$V\left(n, d^{+}(n)\right) = Max\{\Pr\{CS\}\} = \Pr\{CS|S_1\}\Pr\{S_1\} + \Pr\{CS|S_2\}\Pr\{S_2\} + \Pr\{CS|NS\}\Pr\{NS\} \tag{69}$$

$\Pr\{CS \mid S_1\}$ denotes the probability of correct selection when candidate S_1 is selected as the optimal candidate and this probability equals to its belief, $B^+(O_k)$, and with the same discussion, it is concluded that $\Pr\{CS \mid S2\}=1-B^+(O_k)$

$\Pr\{S_1\}$ is the probability of selecting out of control candidate (positive shift) as the solution thus following the decision making strategy, we should have $B^+(O_k)=\max(B^+(O_k),\ 1-B^+(O_k))$ and $B^+(O_k)>d^+(n)$ that is equivalent to following,

$$\Pr\{S_1\}=\Pr\left\{B^+(O_k)>d^+(n)\right\}, d^+(n)\in[0.5,1] \tag{70}$$

With the same reasoning, it is concluded that,

$$\Pr\{S_2\}=\Pr\left\{1-B^+(O_k)>d^+(n)\right\}, d^+(n)\in[0.5,1] \tag{71}$$

1. $\Pr\{CS \mid NS\}$ denotes the probability of correct selection when none of candidates has been selected and it means that the maximum value of the beliefs is less than $d^+(n)$ and the process of decision making continues to latter stage. As a result, in terms of Dynamic Programming Approach, the probability of this event equals to maximum of probability of correct selection in latter stage *(n-1)*, $V(n-1)$, but since taking observations has cost, then the value of this probability in current time is less than its actual value and by using the discounting factor α, it equals $\alpha V(n-1)$
2. Since the entire solution space is partitioned, it is concluded that $\Pr\{CS \mid NS\}=1-(\Pr\{S_1\}+\Pr\{S_2\})$

By the above preliminaries, the function $V(n)$ is determined as follows:

$$\begin{aligned} V(n)= & \\ & \max_{0.5<d^+(n)<1}\left[\begin{array}{l} B^+(O_k)\Pr\left\{B^+(O_k)>d^+(n)\right\}+(1-B^+(O_k))\Pr\left\{1-B^+(O_k)>d^+(n)\right\} \\ +\Pr\{CS|NS\}\left(1-\Pr\left\{B^+(O_k)>d^+(n)\right\}-\Pr\left\{1-B^+(O_k)>d^+(n)\right\}\right) \end{array}\right] \\ = & \max_{0.5<d^+(n)<1}\left[\begin{array}{l} B^+(O_k)\Pr\left\{B^+(O_k)>d^+(n)\right\}+\left(1-B^+(O_k)\right)\Pr\left\{1-B^+(O_k)>d(n)\right\} \\ +\alpha V(n-1)\left(1-\Pr\left\{B^+(O_k)>d^+(n)\right\}-\Pr\left\{1-B^+(O_k)>d^+(n)\right\}\right) \end{array}\right] \end{aligned} \tag{72}$$

In terms of above equation, $V(n,\ d^+(n))$ is obtained as follows:

$$V\left(n,d^+(n)\right)=\left[\begin{array}{l} B^+(O_k)\Pr\left\{B^+(O_k)>d^+(n)\right\}+\left(1-B^+(O_k)\right)\Pr\left\{(1-B^+(O_k))>d^+(n)\right\} \\ +\alpha V(n-1)\left(1-\Pr\left\{B^+(O_k)>d^+(n)\right\}-\Pr\left\{1-B^+(O_k)>d^+(n)\right\}\right) \end{array}\right] \tag{73}$$

Calculation method for $V(n,\ d^{+}(n))$:

$B^{+}(gr,\ O_k)$ and $B^{+}(sm,\ O_k)$ are defined as follows:

$$\begin{aligned} B^{+}(gr,O_k) &= \max\left\{B^{+}(O_k),1-B^{+}(O_k)\right\} \\ B^{-}(gr,O_k) &= \min\left\{B^{+}(O_k),1-B^{+}(O_k)\right\} \end{aligned} \tag{74}$$

Now equation (73) is rewritten as follows:

$$\begin{aligned} &V\left(n,d^{+}(n)\right)=\left(B^{+}(gr,O_k)-\alpha V(n-1)\right)\Pr\left\{B^{+}(gr,O_k)>d^{+}(n)\right\}+ \\ &\left(B^{+}(sm,O_k)-\alpha V(n-1)\right)\Pr\left\{B^{+}(sm,O_k)>d^{+}(n)\right\}+\alpha V(n-1) \end{aligned} \tag{75}$$

There are three conditions:

1. $B^{+}(gr,\ O_k)<\alpha V(n-1)$

In this condition, both $B^{+}(gr,\ O_k)-\alpha V(n-1)$ and $B^{+}(sm,\ O_k)-\alpha V(n-1)$ are negative, thus we should have $d^{+}(n)=1$ in order to maximize $V(n,\ d^{+}(n))$. Since $B^{+}(gr,\ O_k)<d^{+}(n)=1$, we don't select any candidate in this condition and sampling process continues.

2. $B^{+}(sm,\ O_k)>\alpha V(n-1)$

In this condition, both $B^{+}(gr,\ O_k)-\alpha V(n-1)$ and $B^{+}(sm,\ O_k)-\alpha V(n-1)$ are positive, thus we should have $d^{+}(n)=0.5$ in order to maximize $V(n,\ d^{+}(n))$. since $B^{+}(gr,\ O_k)>d^{+}(n)=0.5$, we select the candidate of belief $B^{+}(gr,\ O_k)$ as the solution.

3. $B^{+}(sm,\ O_k)<\alpha V(n-1)<B^{+}(gr,\ O_k)$

In this condition, one of the probabilities in equation (10) has positive coefficient and one has negative coefficient, to maximize $V(n,\ d^{+}(n))$, optimality methods should be applied.

- *Definition:h* $(d^{+}(n))$ is defined as follows:

$$h\left(d^{+}(n)\right)=\frac{d^{+}(n)\left(1-B^{+}(O_{k-1})\right)}{\left(1-d^{+}(n)\right)B^{+}(O_{k-1})} \tag{76}$$

First the value of $\Pr\{B^{+}(O_k)>d^{+}(n)\}$ is determined as follows:

$$\Pr\left\{B^{+}(O_k)>d^{+}(n)\right\}=\Pr\left\{\begin{array}{l}\dfrac{\varphi(x_k)B^{+}(O_{k-1})}{\varphi(x_k)B^{+}(O_{k-1})+\left(1-B^{+}(O_{k-1})\right)0.5}\\ >d^{+}(n)\end{array}\right\}= \tag{77}$$
$$\Pr\left\{\varphi(x_k)>h\left(d^{+}(n)\right)0.5\right\}=1-0.5h\left(d^{+}(n)\right)$$

Since $\varphi(x_k)$ is a cumulative distribution function thus it follows a uniform distribution function in interval [0, 1], thus the above equality is concluded.

With the same reasoning, it is concluded that:

$$\Pr\left\{1-B^{+}(O_k)\geq d^{+}(n)\right\}=\Pr\left\{1-d^{+}(n)\geq B^{+}(O_k)\right\}=0.5h\left(1-d^{+}(n)\right) \tag{78}$$

Now equation (73) can be written as follows:

$$V(n)=\max_{0.5<d^{+}(n)<1}\left[\begin{array}{l}B^{+}(O_k)\left(1-0.5h\left(d^{+}(n)\right)\right)+\left(1-B^{+}(O_k)\right)0.5h\left(1-d^{+}(n)\right)\\ +\alpha V(n-1)\left(1-0.5\left(1-h\left(d^{+}(n)\right)\right)-0.5h\left(1-d^{+}(n)\right)\right)\end{array}\right] \tag{79}$$

And equation (79) can be written as follows:

$$\begin{array}{l}V\left(n,d^{+}(n)\right)=\left(B^{+}(O_k)-\alpha V(n-1)\right)\left(1-h\left(d^{+}(n)\right)0.5\right)+\\ \left(1-B^{+}(O_k)-\alpha V(n-1)\right)0.5h\left(1-d^{+}(n)\right)+\alpha V(n-1)\end{array} \tag{80}$$

Since $V^{*}(n)=\underset{0.5<d^{+}(n)<1}{Max}\left[V\left(n,\ d^{+}(n)\right)\right]$ thus it is sufficient to maximize the real value function $V\left(n,\ d^{+}(n)\right)$, therefore; we should find the function value in points where first derivative is equated to zero as follows,

$$\begin{array}{l}\dfrac{\partial V\left(n,d^{+}(n)\right)}{\partial d^{+}(n)}=0\Rightarrow-\dfrac{\left(B^{+}(O_k)-\alpha V(n-1)\right)}{\left(1-B^{+}(O_k)-\alpha V(n-1)\right)}=\dfrac{\left(1-d^{+}(n)\right)^{2}}{d^{+^{2}}(n)}\\ \Rightarrow d^{+}(n)=\dfrac{1}{\sqrt{-\dfrac{\left(B^{+}(O_k)-\alpha V(n-1)\right)}{\left(1-B^{+}(O_k)-\alpha V(n-1)\right)}}+1}\end{array} \tag{81}$$

The optimal threshold $d^{+}(n)$ is determined by the above equation. Since the optimal value of $d^{+}(n)$ should be in the interval [0.5, 1] thus it is concluded that the optimal value of $d^{+}(n)$ will be determined as follows:

$$d^{+}(n) = Max\left[\frac{1}{\sqrt{-\frac{\left(B^{+}(O_k) - \alpha V(n-1)\right)}{\left(1 - B^{+}(O_k) - \alpha V(n-1)\right)} + 1}}, 0.5\right] \tag{82}$$

The above method is presented for detecting the positive shifts in the process mean and can be adapted for detecting the negative shifts with the same reasoning.

The general decision making algorithm is summarized as follows:

1. Set k=0 and the initial beliefs $B^{+}(O_0)=0.5,\ B^{-}(O_0)=0.5$.
2. Gather an observation and set $k=k+1,\ n=n-1$.
3. If $n<0$, then no shift is occurred in the process mean and decision making stops.
4. Update the values for the beliefs $B^{-}(O_k),\ B^{+}(O_k)$ by equation (61).
5. If $Min\left(B^{+}(O_k),\ 1-B^{+}(O_k)\right)>\alpha V(n-1)$, then if $Max\left(B^{+}(O_k),\ 1-B^{+}(O_k)\right)=B^{+}(O_k)$, it is concluded that a positive shift is occurred in the process mean and decision making stops, also if $Max\left(B^{+}(O_k),\ 1-B^{+}(O_k)\right)=1-B^{+}(O_k)$, then no positive shift is occurred in the process mean and decision making stops.
6. If $Max\left(B^{+}(O_k),\ 1-B^{+}(O_k)\right)<\alpha V(n-1)$, then data is not sufficient for detecting the positive shift and go to stage 2 after checking the occurrence of negative shift in the rest of the algorithm.
7. If $Min\left(B^{-}(O_k),\ 1-B^{-}(O_k)\right)>\alpha V(n-1)$ then if $Max\left(B^{-}(O_k),\ 1-B^{-}(O_k)\right)=B^{-}(O_k)$ it is concluded that a negative shift is occurred the process mean and decision making stops and if $Max\left(B^{-}(O_k),\ 1-B^{-}(O_k)\right)=1-B^{-}(O_k)$, then no negative shift is occurred in the process mean and decision making stops.
8. If $Max\left(B^{-}(O_k),\ 1-B^{-}(O_k)\right)<\alpha V(n-1)$, then data is not sufficient for detecting the negative shift and go to stage 2.
9. If $Max\left(B^{+}(O_k),\ 1-B^{+}(O_k)\right)>\alpha V(n-1)>Min\left(B^{+}(O_k),\ 1-B^{+}(O_k)\right)$, then determine the value of $d^{+}(n)$ (minimum acceptable belief for detecting the positive shift) by the following equation:

$$d^{+}(n) = Max\left(\frac{1}{\sqrt{-\frac{\left(B^{+}(O_k)-\alpha V(n-1)\right)}{\left(1-B^{+}(O_k)-\alpha V(n-1)\right)}+1}}, 0.5\right) \quad (83)$$

10. If $Max\left(B^{-}(O_k),\ 1-B^{-}(O_k)\right)>\alpha V(n-1)>Min\left(B^{-}(O_k),\ 1-B^{-}(O_k)\right)$, then determine the value of $d^{-}(n)$ (minimum acceptable belief for detecting the negative shift) by the following equation:

$$d^{-}(n) = Max\left(\frac{1}{\sqrt{-\frac{\left(B^{-}(O_k)-\alpha V(n-1)\right)}{\left(1-B^{-}(O_k)-\alpha V(n-1)\right)}+1}}, 0.5\right) \quad (84)$$

1. If $B^{+}(O_k)>d^{+}(n)$, then a positive shift is occurred and decision making stops, and if $\left(1-B^{+}(O_k)\right)>d^{+}(n)$, then no positive shift is occurred and decision making stops, else go to stage 2 after checking the occurrence of negative shift in rest of the algorithm.
2. If $B^{-}(O_k)>d^{-}(n)$, then a negative shift is occurred and decision making stops, and If $\left(1-B^{-}(O_k)\right)>d^{-}(n)$, then no negative shift is occurred and decision making stops, else go to stage 2.
3. The approximate value of $\alpha V(n-1)$ based on the discount factor α in the stochastic dynamic programming approach is $\alpha^{n} V(0)$.

5. Conclusion

In this chapter, we introduced a new approach to determine the best solution out of m candidates. To do this, first, we defined the belief of selecting the best solution and explained how to model the problem by the Bayesian analysis approach. Second, we clarified the approach by which we improved the beliefs, and proved that it converges to detect the best solution. Next, we proposed a decision-making strategy using dynamic programming approach in which there were a limited number of decision-making stages.

http://dx.doi.org/10.5772/57423

Appendix 1

Conditional Mean and Variance of the Variables

Conditional mean of variables gr and sm can be evaluated using the following equation.

$$\left(\mu_{sm}, \mu_{gr} \mid (\mu_j)_{j\neq gr,sm}\right)=\left(\mu_{sm}, \mu_{gr}\right)+\boldsymbol{b}_2'\left((\boldsymbol{X}_{kj})_{j\neq gr,sm}-(\boldsymbol{\mu}_j)_{j\neq gr,sm}\right) \tag{85}$$

where, $\boldsymbol{b}_2'=\boldsymbol{\Sigma}_{xX}\boldsymbol{\Sigma}_{XX}^{-1}$

and

$$\boldsymbol{\Sigma}=\begin{bmatrix}\boldsymbol{\Sigma}_{XX} & \boldsymbol{\Sigma}_{xX}\\ \boldsymbol{\Sigma}_{xX} & \boldsymbol{\Sigma}_{xx}\end{bmatrix} \tag{86}$$

$\boldsymbol{\Sigma}$: The covariance matrix of the process

$\boldsymbol{\Sigma}_{xx}$: Submatrix of the covariance matrix $\boldsymbol{\Sigma}$ for variables $j=gr,\ sm$

$\boldsymbol{\Sigma}_{xX}$: Submatrix of the covariance matrix $\boldsymbol{\Sigma}$ between variables $j=gr,\ sm$ and $j\neq gr,\ sm$

$\boldsymbol{\Sigma}_{XX}$: Submatrix of the covariance matrix $\boldsymbol{\Sigma}$ for variables $j\neq gr,\ sm$

Further, the conditional covariance matrix of variables $j=gr,\ sm$ on variables $j\neq gr,\ sm$, is obtained as $\boldsymbol{\Sigma}_{xx}-\boldsymbol{\Sigma}_{xX}^{T}\boldsymbol{\Sigma}_{XX}^{-1}\boldsymbol{\Sigma}_{xX}$.

Appendix 2

Evaluating the Optimal Value of $d_{gr,sm}(k)$

Assume $(\mu_j)_{j\in\{1,2,\ldots,m\}}=0$ and $(\sigma_j)_{j\in\{1,2,\ldots,m\}}=1$. Then,

$$\begin{aligned}&\Pr\left\{\left\{e^{0.5(T_{k,gr|sm})^2}\geq h\left(d_{gr,sm}(k)\right)e^{0.5(T_{k,sm})^2}\right\}\right\}=\\ &\Pr\left\{0.5(T_{k,gr|sm})^2\geq \ln\left(h\left(d_{gr,sm}(k)\right)\right)+0.5(T_{k,sm})^2\right\}=\\ &\Pr\left\{(T_{k,gr|sm})^2-(T_{k,sm})^2\geq 2\ln\left(h\left(d_{gr,sm}(k)\right)\right)\right\}\end{aligned} \tag{87}$$

Now since $(T_{k,sm}, T_{k,gr\mid sm})$ follow a standard normal distribution $(\mu_j)_{j\in\{gr,sm\}}=0$ and $(\sigma_j)_{j\in\{gr,sm\}}=1$, hence $(T_{k,gr\mid sm})^2$ and $(T_{k,sm})^2$ follow a χ^2 distribution with one degree of freedom. Then using an approximation, if we assume that $(T_{k,sm})^2$ is approximately equal to its mean, we have

$$(T_{k,sm})^2 \propto E\left(T_{k,sm}{}^2\right) = E\left(T_{k,sm}\right)^2 + Var\left(T_{k,sm}\right) = 1 \tag{88}$$

Thus,

$$\begin{aligned}
&\Pr\left\{(T_{k,gr|sm})^2 - (T_{k,sm})^2 \geq 2\ln\left(h\left(d_{gr,sm}(k)\right)\right)\right\} \propto \\
&\Pr\left\{(T_{k,gr|sm})^2 - E(T_{k,sm}{}^2) \geq 2\ln\left(h\left(d_{gr,sm}(k)\right)\right)\right\} \\
&= \Pr\left\{(T_{k,gr|sm})^2 - 1 \geq 2\ln\left(h\left(d_{gr,sm}(k)\right)\right)\right\} = \\
&\Pr\left\{(T_{k,gr|sm})^2 \geq 2\ln\left(h\left(d_{gr,sm}(k)\right)\right) + 1\right\}
\end{aligned} \tag{89}$$

Now, since $(T_{k,gr|sm})^2 \propto \chi^2(1)$, we have

$$\Pr\left\{(T_{k,gr|sm})^2 \geq 2\ln\left(h\left(d_{gr,sm}(k)\right)\right) + 1\right\} = \int_{2\ln\left(h\left(d_{gr,sm}(k)\right)\right)+1}^{\infty} \frac{e^{-\frac{t}{2}} t^{-\frac{1}{2}}}{\Gamma\left(\frac{1}{2}\right) 2^{\frac{1}{2}}} dt \tag{90}$$

Hence,

$$\Pr\left\{(T_{k,gr|sm})^2 - (T_{k,sm})^2 \geq 2\ln\left(h\left(d_{gr,sm}(k)\right)\right)\right\} \simeq \int_{2\ln\left(h\left(d_{gr,sm}(k)\right)\right)+1}^{\infty} \frac{e^{-\frac{t}{2}} t^{-\frac{1}{2}}}{\Gamma\left(\frac{1}{2}\right) 2^{\frac{1}{2}}} dt \tag{91}$$

Similarly

$$\begin{aligned}
&\Pr\left\{B_{gr,sm}\left(gr;O_k\right) \geq d_{gr,sm}(k)\right\} = \Pr\left\{e^{0.5(T_{k,gr|sm})^2} \geq r\left(d_{gr,sm}(k)\right) e^{0.5(T_{k,sm})^2}\right\} \simeq \\
&\int_{2\ln\left(r\left(d_{gr,sm}(k)\right)\right)+1}^{\infty} \frac{e^{-\frac{t}{2}} t^{-\frac{1}{2}}}{\Gamma\left(\frac{1}{2}\right) 2^{\frac{1}{2}}} dt
\end{aligned} \tag{92}$$

Replacing the above equations in equation (53) results in

$$\begin{aligned}
V_{i,j}{}^{*}(N) \propto &\left(B_{gr,sm}\left(gr;O_k\right) - \alpha V_{i,j}^{*}\left(N-1\right)\right) \int_{\left(2\ln\left(r\left(d_{gr,sm}(k)\right)\right)+1\right)}^{\infty} \frac{e^{-\frac{t}{2}} t^{-\frac{1}{2}}}{\Gamma\left(\frac{1}{2}\right) 2^{\frac{1}{2}}} dt + \\
&\left(B_{gr,sm}\left(sm;O_k\right) - \alpha V_{i,j}^{*}\left(N-1\right)\right) \int_{\left(2\ln\left(h\left(d_{gr,sm}(k)\right)\right)+1\right)}^{\infty} \frac{e^{-\frac{t}{2}} t^{-\frac{1}{2}}}{\Gamma\left(\frac{1}{2}\right) 2^{\frac{1}{2}}} dt + \alpha V_{i,j}^{*}\left(N-1\right)
\end{aligned} \tag{93}$$

http://dx.doi.org/10.5772/57423

Now by solving the equation $\frac{\delta V_{i,j}(N)}{\delta d_{gr,sm}(k)}=0$, the following equation is obtained.

$$\left(B_{gr,sm}(gr;O_k)-\alpha V^*_{gr,sm}(N-1)\right)\frac{1}{\sqrt{\left(\ln\left(r\left(d_{gr,sm}(k)\right)\right)+1\right)}}= -\left(B_{gr,sm}(sm;O_k)-\alpha V^*_{gr,sm}(N-1)\right)\frac{1}{\sqrt{\left(\ln\left(h\left(d_{gr,sm}(k)\right)\right)+1\right)}} \tag{94}$$

Finally, the approximate value of $d_{gr,sm}(k)$ say $d^1_{gr,sm}(k)$ is determined by solving this equation numerically or by a search algorithm.

Now using another approximation, if we assume that $(T_{k,gr})^2$ is approximately equal to its mean, the approximate value of $d_{gr,sm}(k)$ say $d^2_{gr,sm}(k)$ is determined by solving following equation,

$$\left(B_{gr,sm}(sm;O_k)-\alpha V^*_{gr,sm}(N-1)\right)\frac{1}{\sqrt{\left(\ln\left(r\left(d_{gr,sm}(k)\right)\right)+1\right)}}= -\left(B_{gr,sm}(gr;O_k)-\alpha V^*_{gr,sm}(N-1)\right)\frac{1}{\sqrt{\left(\ln\left(h\left(d_{gr,sm}(k)\right)\right)+1\right)}} \tag{95}$$

The approximate optimal value of $d_{gr,sm}(k)$ is obtained as follows,

$$d_{gr,sm}(k)=\max\left\{d^1{}_{gr,sm}(k),d^2{}_{gr,sm}(k)\right\} \tag{96}$$

Author details

Mohammad Saber Fallah Nezhad*

Address all correspondence to: Fallahnezhad@yazd.ac.ir

Associate Professor of Industrial Engineering, Yazd University, Iran

References

[1] Basseville, M. & Nikiforov, I.V. (1993). Detection of abrupt changes: Theory and application. (Information and System Sciences Series), Prentice-Hall.

[2] Eshragh-J., A. & Modarres, M. (2009). A New Approach to Distribution Fitting: Decision on Beliefs, Journal of Industrial and Systems Engineering, 3(1), 56-71.

[3] Saniee Monfared M.A., Ranaeifar F. (2007), Further Analysis and Developments of DOB Algorithm on Distribution Fitting; Scientia Iranica, 14(5), 425-434.

[4] Eshragh-J., A. & Niaki, S.T.A. (2006). The application of decision on beliefs in response surface methodology. Proceedings of the 4th International Industrial Engineering Conference, Tehran, Iran.

[5] Bechhofer, R.E. & Kulkarni, R. (1982). On the performance characteristics of closed adaptive sequential procedures for selecting the best *Bernoulli* populations. *Communication in Statistics – Sequential Analysis,* 1, 315-354.

[6] Gupta, S.S. & Panchapakesan, S. (2002). Multiple decision procedures: Theory and methodology of selecting and ranking populations. Cambridge University Press, U.K.

[7] Nair, V.N., Tang, B., and Xu, L. (2001). Bayesian inference for some mixture problems in quality and reliability, Journal of Quality Technology, 33, 16-28.

[8] Niaki, S.T.A., Fallahnezhad, M. (2009), Decision-making in detecting and diagnosing faults in multivariate statistical quality control environments. International Journal of Advanced Manufacturing Technology 42, 713-724

[9] Mason, R.L., Tracy, N.D., Young, J.C. (1995).Decomposition of T^2 for multivariate control chart interpretation. *Journal of Quality Technology* 27, 99-108.